SUSTAINABLE TRAVEL

The essential guide to positive-impact adventures

HOLLY TUPPEN

Contents

Introduction

I fell into sustainable travel off the back of an adventure. My boyfriend and I bought a world map and marked anything that intrigued us. One day we drew a line between them, and the trip suddenly sprung to life; we decided to circumnavigate the world without flying. Before we knew it, ferries turned into sailing expeditions, buses into 3,000-mile cycling adventures and the itinerary filled up with unknown places. The trip has had a lasting impact on everything I think, from love to politics.

On returning to London, I was lucky enough to land some dream jobs: working for no-fly platform Green Traveller, as Editor of *Green Hotelier*, and as a sustainable travel expert and writer for the likes of *The Long Run*, the *Guardian* and the World Travel and Tourism Council. Slow travel morphed into sustainable travel, which soon hit the mainstream.

But the sweet spot didn't last for long. In recent years, a steady tension has developed between the travel that I love and its uglier side. On the one hand, it brings inspiration, creates connections, and has the power to renew people and places. On the other, it pushes locals out, causes environmental damage, and has become an unrelenting commodity.

I've been enraged by greenwashing, devastated by mounds of plastic, and heartbroken on hearing about lives destroyed by climate change and overtourism. The more I've learned about the climate crisis, and the deeper I've plunged into the commercial side the travel, the harder it's been to stomach the hypocrisy of travelling sustainably.

Meanwhile, during countless press trips, adventures and interviews, I've been overwhelmed by the power of sustainable travel. I've shed tears of awe at how people can spend their lives striving for positive change despite the odds, and tears of wonder at the complexity, beauty and magnitude of humanity's connection to the natural world. In grinding the world to an unimaginable standstill, the COVID-19 crisis has crystallized all that is good and bad about our travel habits.

Rather than battle it, I've come to accept this lurching between the highs and lows of travel. It keeps me in check and encourages me to challenge. It stops me greedily jumping on a plane at every opportunity, and helps me to embrace doorstep adventures. When I do get on a plane, it drives me to make it count, by supporting the changemakers and telling their story; since tourism was knocked sideways by COVID-19, this is more important than ever.

I hope this book will take you on a similar journey. First, exploring why we're talking about sustainability and why we need to act. Next, what that action looks like; how we can all make more informed choices. There are some pointers on how to identify genuine sustainable travel experiences and develop 'responsible travel intuition' to cut through the greenwash. Next comes the inspiration. In 'Regenerative travel' (pages 70–113), we pour over examples of how travel can change the world for good – whether protecting heritage, supporting communities or saving a species. Finally, there is a whirlwind guide to making travel count in each continent.

Unlike so many travel books, it's not all glossy. When it comes to travelling responsibly, there's no quick win, so there are some hefty questions raised throughout. Sustainability is complex, and we need open debate and mindful actions to progress.

But we mustn't be discouraged. As we dive into a new decade – the most critical decade we've ever known for the survival of our species – the travel industry is teetering on precipice. It can choose the long-term sustainable path, or it can plummet down the self-destructive one. As holiday-goers, explorers, campaigners, voters, adventurers and people who draw energy from the Earth's natural wonders and cultural diversity, we can collectively guide it towards the light.

Can we travel better?

—

To understand why we need to travel more sustainably, as well as how to do so, we first need to unravel the increasingly complex question: can we justify travel in a climate crisis? It sounds heavy, and it would be impossible to answer in full without exploring the scientific context, but the purpose of this book is to prove that we can justify travel, but only if we change our ways.

This chapter explores why this question is being asked, why we should care (it's a holiday after all) and lays the foundations of a sustainable travel mindset. Being a more conscientious traveller is not only for the benefit of the planet, nor simply for bragging rights, but it is essential if we want to continue to do what we love: explore and understand the world around us.

What is sustainable travel?

The United Nations agency, the World Tourism Organization, defines sustainable tourism as 'tourism that takes full account of its current and future economic, social and environmental impacts, addressing the needs of visitors, the industry, the environment and host communities'.

But such a vague definition is no longer fit for purpose. Surely the needs of visitors and the travel industry should come after the needs of the environment and host communities? Is now the time to move beyond taking full account of impact, and instead have a positive impact wherever we go?

Rather than dwelling on the notion of sustainability (where sustainability simply means being able to continue over and over again), perhaps now, in the midst of an urgent environmental crisis, we need to seek out transformative and regenerative experiences: travel that funds the protection of nature; travel that empowers marginalized parts of society; travel that is designed and driven by communities; travel that helps us learn from indigenous peoples; travel that heals; travel that brings us closer to our doorstep natural and cultural heritage; travel that uplifts the host as much as the visitor; travel that changes hearts and minds for the better.

When we start to think like this, rather than celebrate reaching the pinnacle of responsible travel, it becomes clear we're only just getting started. How can we travel better on home turf? How do we democratize travel? What about virtual travel? How can travel better champion diversity and inclusivity? How can travel make places more resilient to climate change? As travellers, we start to look inward, too: how can our travel experiences help us connect more profoundly with the natural world and each other?

In this momentum, it's evident that regardless of terminology, sustainable travel is a journey. It's not something we're going to achieve or tick off; we just have to keep learning, adapting and inspiring others to do the same. The quicker we all jump on board and rise to the challenge, the better.

The climate and biodiversity crisis

The word 'sustainability' has become such an everyday part of our vocabulary that it's easy to forget why we're talking about it. So before we delve into how travel can do good, let's rewind in order to better understand why we should care.

Climate change is an issue that's been simmering since the 1970s. In primary school, thirty-odd years ago, I learned about deforestation, rising temperatures and animal extinctions. I remember it acutely because it was so unbelievable to me: why would humans do this? Rather depressingly, my kids are now going through the same torment of emotions over the same issues. For a species so quick to advance and progress, we're certainly taking our time when it comes to the very thing that could wipe us out.

It's no wonder that we have felt confused, unable to discern the truth or to work out the best course of action to take. Some of us have buried our heads in the sand, while a self-defeating sense of hopelessness cripples others.

However, in 2019 the climate debate became headline news and a 'crisis' was born. Many say it's thanks to prominent spokespeople like Greta Thunberg, who rocketed to global fame with the help of social media and a purpose-hungry youth. Others say it's a scapegoat for general political and social malaise – an expression of angst with nowhere else to turn. But, in reality, climate change became a climate crisis in 2019 because the nightmarish predictions became a living reality.

Rising temperatures have started to change weather patterns and impact lives. Wildfires in California, bushfires in Australia, Hurricane Maria ravaging Puerto Rico and two consecutive cyclones in Mozambique – a region never before hit by cyclones – are just some of the devasting climate events in recent years. We're also now experiencing the slow creep of rising temperatures making cities unliveable, causing year-round drought, and putting more than 8.6 million Americans at risk from coastal flooding.

Of course, the Earth has experienced climate fluctuations in the past. For decades, this argument, propelled by market forces, fossil fuel lobbyists and short-term political systems, has made it easy to dismiss concerns and continue as normal, which means churning out more carbon and destroying more ecosystems to create more stuff, more convenience and more comfort for less money. Short-term gain has prevailed over long-term prosperity.

But this time it's different. The global average of atmospheric carbon dioxide is now approximately 407.4 parts per million – an increase of 25 per cent since we started burning fossil fuels on a mass scale in the 1950s. The change is creating temperatures unknown on Earth since prehistoric times.

Alongside our carbon problem, we're experiencing a biodiversity crisis. The continued destruction of habitats (for farming, mining or urban sprawl) means that we're losing species at a thousand times their natural rate of decline. The 2020 Living Planet Index shows that wildlife species have declined by 68 per cent since 1970, and we've lost natural habitats eight times the size of the UK since 2000. In short, we've picked, logged, fished, hunted, packed, mined and plucked to excess. The World Wildlife Fund predicts that we're currently using 25 per cent more natural resources than the planet can sustain.

Unlike previous climate and biodiversity fluctuations of this scale, the crisis we're experiencing now is humanmade. So devastating is our biological, chemical and physical impact on Earth that scientists believe we're entering a new geological age. For the first time in Earth's history, humans are the driving force of environmental change. The new epoch is known as the Anthropocene.

It sounds damning, but there is hope. A *Nature* report in 2020 used modelling to suggest that we can halt or even reverse habitat loss and deforestation if we act now, particularly by changing our eating habits. Commenting on the 2020 Living Index report, Sir David Attenborough said that this could be the decade we become genuine stewards of our planet. 'Doing so will require systemic shifts in how we produce food, create energy, manage our oceans and use materials,' he told the BBC. 'But above all it will require a change in perspective. A change from viewing nature as something that's optional or "nice to have" to the single greatest ally we have in restoring balance to our world.'

WILL YOU JOIN
GRETA?
SKOLSTREJK
FÖR
KLIMATET
AGENT
PETRUSCIONI

Travel in the Anthropocene

If there's one question rocking our wanderlust these days, it's this: is it morally acceptable to travel the world for fun while it burns up around us?

The problem is that tourism infrastructure can damage wildlife and habitats, and our jet-setting habits contribute to a fair amount of carbon emissions. The tourism industry is responsible for between 8 to 12 per cent of the world's carbon emissions. That's less than the fashion industry, but more than construction. Although it might not sound much, it's the industry's pre-COVID-19 growth that has rung alarm bells.

As countries like China get wealthier, more people, naturally, want to travel. In 2019, international arrivals reached 1.4 billion, two years ahead of schedule (according to the World Tourism Organization). That's almost 20 per cent of the world's population on the move. If we return to this trajectory, the increase in the demand for tourism will outstrip efforts to decarbonize the industry.

These statistics throw some traditional justifications for sustainable travel off-kilter. Will we only be moved to protect the world if we experience it? Do we need to fly to the other side of the world to become global-minded citizens? Can we only understand climate change if we see its impact first-hand?

While these statements are valid considerations, it's worth noting that if 7.7 billion people need to travel the world before taking environmental action, we're in serious trouble.

The COVID-19 global lockdown proved that there are other solutions. Not only did more of us embrace doorstep adventures, but we learnt to communicate with all four corners of the world from our sofas. We ventured into the bush on virtual safaris, got up-close to world-famous art, explored architecture with local experts, and even trained to become marine biologists. We achieved all this while burning considerably less fossil fuels. Experiences that were once reserved for the few became available to the many.

DELPHINE MALLERET KING,
MANAGING DIRECTOR, THE LONG RUN

The Long Run brings together some of the world's most committed nature-based travel businesses operating according to the 4Cs: Conservation, Community, Culture and Commerce.

–

What's next for sustainable travel?

We've always worked with members to embed sustainability into their operations and business strategy. This is now happening across the board, from global tour operators to tiny lodges. This is a vital next step, because until sustainability becomes a way of doing business and the 4Cs becomes a decision-making process, it will remain an add-on, and innovative thinking and positive impact will be limited.

Also, we are all growing more mindful of how and why we travel, and more people in the industry will hopefully harness this to drive impact. How can travel better connect us to nature and to each other? How can travel contribute further to touch individuals and transform society for the better?

What gives you most hope for a more sustainable future?

The world is waking up to the fact we are in a biodiversity and climate crisis, and that everyone has a part to play in overcoming them. Consumers are looking for more purposeful, low-impact experiences, and businesses are starting to think more holistically – recognizing that sustainability isn't something that can just happen on the sidelines. The young are the most motivated, which brings a huge amount of hope for systemic change in how business operates.
I believe that we are at the cusp of a significant change – uncertainty and unpredictability will hopefully encourage more creativity and out-of-the-box thinking.

What advice can you offer to those looking to travel more sustainably?

I would urge anyone to use The Long Run's 4Cs as a framework to consider the negative or positive impact of a trip – is it supporting or uplifting communities, culture, conservation and commerce? Where possible, travel less frequently but for longer, mindfully and meaningfully. Think about your carbon and social footprint; less can be more. And finally, seek experiences that have a tangible positive impact – whether protecting a forest, keeping a species alive or fostering exchange, understanding and tolerance.

Of course, even the most expert production values can't replace being there in person, but the accessibility of the world online does provide another reason to pick and choose our real-life experiences even more carefully. If we continue to travel in the Anthropocene, we need to make every trip count. This means travelling less, travelling for longer, and travelling better.

A useful framework to have in mind is the UN's 17 Sustainable Development Goals. Adopted in 2015, these goals provide a blueprint for 'peace and prosperity for people and the planet, now and into the future'. It's an urgent call of action for countries, but can equally be applied to businesses and individuals. The framework highlights that peace, the environment, and poverty alleviation are connected and depend on collective action.

Another helpful blueprint comes from the conservation-led travel not-for-profit The Long Run, founded by Jochen Zeitz. The Long Run's 4Cs – a balance of Conservation, Community, Culture and Commerce – help lodges embed sustainable thinking into everything they do. So far, the model has helped to protect over 20 million acres of ecosystems and improve the lives of over 750,000 people.

Whether travelling to a Long Run destination or not, 4C thinking can direct us to experiences with more positive impact. When contemplating a trip, we need to ask ourselves: is it helping to conserve nature, support a local community and protect culture in a financially sustainable and low-impact way?

Genuinely sustainable travel inspires us to become caretakers of the natural world, but it doesn't stop there. It is also solution-focused; it improves destinations, uplifts communities, and provides a financial incentive to protect landscapes, wildlife and culture. It's also more rewarding for the traveller; by seeking experiences that actively do good, then the chances are we'll be more fulfilled as travellers and holiday-goers, too. Any experience built on a genuine love for a place, landscape or community will always surpass those that don't. The alternative – that our jet-setting habits destroy the very things that inspire our wanderlust – surely isn't worth the risk.

Understanding carbon

To travel sustainably, we need to halt the climate crisis. To do that we need to reduce carbon emissions – the earth-choking, atmosphere-warming particles that are released when we burn the fossil fuels that help us manufacture products, generate power and make our world more efficient and comfortable.

Scientists in the know, from the Intergovernmental Panel on Climate Change (IPCC), state that a 1.5°C increase in global warming is the limit the planet can handle, and staying below this rise means mitigating the devastating impact on the world and its inhabitants. In some places – like the Arctic – we've already exceeded this temperature rise, in others, we're not yet close. That's because global warming as a result of burning fossil fuels is not uniform around the world.

If we don't respect this upper temperature limit across the planet, things will quickly move from dire to catastrophic. If we reach 3°C of global warming in the next century, sea levels will rise by 3 feet, displacing 680 million people in low-lying coastal zones, along with 65 million citizens of small island states. That's not to mention the fires, extreme weather events and drought.

At the current trajectory, it is estimated that we will reach the 1.5°C limit as early as 2030. To halt our advance beyond it, we need to globally cut carbon emissions in half by then.

So, what's this got to do with travel? Everything we consume is responsible for a proportion of carbon emissions and our holiday plans are no exception.

The impact of travel

Our travel choices can have a significant impact on our individual carbon footprint. A 2017 report conducted by Lund University in Sweden found that a round-trip to Australia creates 4 tonnes of carbon emissions – that's twenty times more than the carbon saving made by a year of recycling. It blows what the World Resources Institute deems to be a responsible personal carbon allowance (2.5 tonnes per year) out of the water.

The plot thickens when we consider findings from Lenzen *et al.*'s *Nature Climate Change* study published in 2018. The report found that 'growth in tourism is a stronger accelerator of carbon emissions than the growth in manufacturing, construction and services provision'. Although tourism provides one in ten jobs and 10 per cent of global GDP, when it comes to carbon, it's currently one of the least efficient ways to create economic value.

So to travel sustainably, we need to think about carbon every step of the way. Flying often bears the brunt of this discussion, and rightfully so given it's the single most polluting thing we can do as an individual (after having children), but flying to your destination doesn't mean that all is lost.

A carbon footprint report commissioned by Responsible Travel in 2020 found that by making low-carbon choices while on holiday, our carbon emissions can meet the global sustainable average per day. It says that 'where more climate-friendly choices are made (food, transport and accommodation), emissions for a holiday can be very close to the global sustainable average per day (10 kg CO_2-e) and almost half the current average per day, per person emissions in the UK (20 kg CO_2-e)'.

This is good news. Rather than having to feel guilty every time we follow our wanderlust, we simply need to make low-carbon choices. By doing so, we could even get our travel carbon footprint lower than our footprint at home. These choices might include a plant-based diet, using public transport and choosing low-impact, energy-saving places to stay.

Lufthansa
Lufthansa
Lufthansa
lufthansa.com
M
M

Reducing your carbon footprint

There are plenty of ways to reduce your carbon footprint on holiday, and while it's unlikely you can tick every box, it's better to tick one than none at all. Those with more time on their hands might choose to travel slow, others might opt for a climate-friendly diet or low-impact accommodation. Either way, here are several ways to minimize emissions when travelling.

Fly less

While boarding an easyJet flight to Sweden last year, I started to feel uneasy. On take-off, tears began to roll down my cheeks and the usually awe-inducing sight of the River Thames sparkling beneath a clear sky couldn't lift my guilty mood. By the time we reached Gothenburg, I was a nervous wreck. It was the first time I had felt the physical effects of eco-anxiety.

We can't talk about carbon and travel without addressing the elephant in the room – aviation. The aviation industry emits more carbon in one year than the whole of Germany. In 2019, despite launching the (now banned by regulators) 'greenest airline in Europe' campaign, Ryanair was Europe's ninth biggest polluter. The top eight were coal mines.

Part of the problem is that the airline industry is heavily subsidized. According to Responsible Travel's campaign for Green Flying Duty, unlike car petrol, international aviation fuel is exempt from tax and VAT. This makes flying up to four times cheaper than it would be otherwise. It also means that those who don't fly (the vast majority of the global population), often end up paying the price for something they didn't do.

Although only responsible for 2.5 per cent of the world's carbon emissions, it is aviation's growth that presents the problem. In the last five years, carbon emissions from aviation have shot up 32 per cent, and yet still only 5 per cent of the world's population flies. In terms of generating demand for oil, aviation is up there with passenger cars. Climate experts also agree that the negative impact of aviation is multiplied by other toxic gases (including nitrogen oxide) and altitude.

Recognizing the problem, airlines and aerospace firms are trying to turn things around. Biofuel (greener than jet fuel) is already in use, while electric planes are on the way, but both are

unlikely to curb emissions fast enough to halt irreversible damage to the planet.

The International Energy Agency states that 'aviation biofuel production of about 15 million litres in 2018 accounted for less than 0.1 per cent of total aviation fuel consumption'. Short-haul and medium-range electric planes, or dual electrical and conventional fuel planes, are expected to be ready within the next thirty years. The Green Flying Duty would help to fund this innovation.

In the meantime, flying comes with a hefty carbon footprint. To put it in context, a return flight from London to New York emits more carbon than an average person does in a whole year in most African countries. So if you want to get serious about reducing your carbon emissions while travelling, the best place to start is to reduce flying habits.

Choose alternative transport

Ditching air travel doesn't mean forgoing adventures and holidays – quite the opposite. Going slow and jumping on a train, boat or bike adds a whole new dimension of discovery to any travel plan. Read more about this in 'Planning a Trip' (pages 32–69).

It's also easier than ever. Historically, the travel industry has been dependent on flights, but change is coming. Tour operators are adding flight-free itineraries to their books, while some governments are taking the lead in investing in better coach, bus, train and ferry infrastructures. Dutch airline KLM has even replaced one of its domestic flights with a high-speed rail link.

Although there's no question that flying is the least eco-friendly way to travel, there are some fluctuations among the next in rank.

How green a train, ferry, coach or car is depends on how many people are in it and what fuel or

JUSTIN FRANCIS, CHIEF EXECUTIVE OFFICER, RESPONSIBLE TRAVEL

Responsible Travel is an activist travel company committed to making tourism a more caring industry and connecting travellers with over 400 specialist and sustainable travel companies.

–

What's next for sustainable travel?

For tourism to be a force for good we need to contribute to addressing what I see as the three biggest global challenges facing mankind: the climate crisis, biodiversity loss and poverty. In a time of crisis anything else is distraction. Specifically, we need to do three things by 2030: fly 30 per cent less over the next ten years to limit the rise in average global temperatures to 1.5°C above pre-industrial levels; deliver a biodiversity net gain in all our businesses; contribute to the UN's sustainable development goal to eliminate poverty.

What gives you most hope for a more sustainable future?

Firstly, there is the growing realization that we need a fair taxation of aviation (if aviation fuel was taxed as petrol is then flying would be up to four times as expensive) and ideas like a frequent flyer levy to make this more socially just. Secondly, destinations are increasingly standing up to the tourism industry, demanding 'tourism on our terms' and regulating it. Lastly, the pioneering work of many smaller responsible tourism businesses over the past thirty years – especially around biodiversity and inequality – is starting to be adopted by the biggest tourism companies.

What advice can you offer those looking to travel more sustainably?

Make positive choices. Plan a holiday that improves local lives and conserves wildlife, habitats and cultural heritage. And be sure to enjoy yourself – you'll have a more authentic experience. Access to authenticity is a gift given at the discretion of local people who feel positive about tourism and tourists.

power source is used. For example, Eurostar, the high-speed rail link between London and Paris, has a much lower carbon footprint than other rail services because more than 50 per cent of electricity in France and England is generated from renewable energy sources.

Unless we all walk everywhere, there's no perfect solution, but it's helpful to be informed to make your own decision. Here are some stats that paint a clearer picture:

- According to US non-profit organization the Union of Concerned Scientists, the average CO_2 emissions from a bus is 0.17 lb/passenger mile, from a train it is 0.41 lb/passenger mile, from a car it is 1.17 lb/passenger mile, and from an airplane it is 1.83 lb/passenger mile.
- Catching the Eurostar from London to Paris emits a staggering 90 per cent fewer carbon emissions than flying the same distance. Electric trains tend to have a 20 to 30 per cent lower carbon footprint than diesel trains.
- Mile for mile, driving a diesel car is not that much better than flying. However, once you get four passengers in the car, the carbon emissions per person can shrink to a third of flying. It's better still when you take into account the increased warming effect and other pollutants of a plane at altitude.
- To enjoy the convenience of driving but with fewer emissions, electric is the way to go. Consider hiring a vehicle – this is more practical in some countries than others. Amsterdam has the highest concentration of EV (electric vehicle) charging stations, while Norway has the highest number of electric cars on the road.
- Statistics from ferry companies on carbon emissions are vague, but the UK government estimates that ferry travel has one third the carbon footprint of flying per passenger.
- Real eco-halos are reserved for those that travel by their own steam – sailing, cycling, walking, kayaking, paddle-boarding or whatever other wonderful way takes your fancy that doesn't burn a single jot of fuel.

Go for longer

Rather than taking several short trips a year, it's more carbon-efficient to fly to one destination for a more extended trip. If you do take shorter trips, try to travel by means other than airplane.

Responsible Travel's 2020 carbon footprint report proved that the gap between transport and other emissions sources, such as food and accommodation, begins to close when a traveller increases their length of stay. So, by going for longer, you can reduce your per-day holiday carbon footprint.

Perhaps leading the curve in a shift towards longer trips, Netherlands-based tour operator Better Places provides a minimum of seventeen-day itineraries. It has also reduced the number of stops in one trip from eight to five, to reduce in-destination travel.

Fly better

If you do have to fly, there are some simple ways to reduce emissions:

- Emissions are highest on take-off and landing, so fly direct and avoid layovers.
- Per mile, short-haul flights emit more carbon, so it makes sense to cut back on flights for weekend breaks.
- If you're going long-haul, resist hopping around in countries or continents by plane and look for other transport options.
- Airplanes emit water vapour during flights that can cause the formation of ice clouds, or 'contrails'. These can lead to a net warming factor estimated to be 2.7 times the typical effect. Fly during the day to reduce the effect of contrails, which is doubled at night because contrails trap heat from Earth, but don't reflect the sun's rays into space.
- Stick to economy. Luxuriating in business and first class can quadruple your carbon footprint because you take up more space.

- Choose an efficient airline that is committed to reducing carbon emissions (without relying on offsetting – see page 31) and is investing in cleaner fuel and green technology. Note that modern aircrafts tend to be more efficient.
- Doing your research and uncovering the truth about airlines and carbon emissions is not easy, but there are a few handy tools. German carbon consultancy, Atmosfair, ranks airlines for their carbon emissions each year. A recent report found that the most efficient models are Boeing 787-9, Airbus A350-900 or the A320neo, and efficient airlines include TUI Airways and TUIfly, Condor, LATAM, Air Europa and KLM. Use Atmosfair's online flight emissions calculator to compare the carbon emissions of specific flights and journeys.

Carbon beyond transport

With *flyskam* (the Swedish word for feeling flight-shame) getting so much attention in recent years, it's easy to forget that we should feel responsible for reducing our emissions while we're on holiday, too. Although transport accounts for the bulk of carbon emissions, where we stay, what we eat and how we get around in a destination also has a considerable impact.

The good news is that reducing your holiday carbon footprint isn't rocket science. Considering that 'stuff' has a carbon footprint just like we do, the more goods – food, interiors, construction materials etc. – that have to be transported for the sake of your holiday, the higher your carbon footprint. Going local wherever possible is therefore an essential part of decarbonizing your travel, and that includes choosing hotels built using local materials to indulging in locally sourced and seasonal food.

STOP AND THINK: CARBON-NEUTRAL

The rising tide of concern about the climate crisis has created a wave of claims, from carbon-neutral around-the-world trips to carbon-positive hotels. One minute Hawaii is announcing plans to become a carbon-neutral country, and the next Swedish burger joint, Max Burger, serves up the world's first climate-positive burger. Deciphering what all this means, which claims stand up and which are greenwashing, can be a challenge.

'Carbon-neutral' refers to an activity or product that releases net-zero carbon emissions into the atmosphere. 'Climate-positive', meanwhile, goes beyond this to create environmental *benefit* by removing additional carbon from the atmosphere. This is also sometimes known as 'carbon-negative' or 'carbon-positive'.

Each of these terms and claims needs scrutiny. If a trip, hotel or airline has achieved carbon neutrality purely through offsetting its emissions, it's not addressing the real issue. Reducing absolute carbon emissions (by changing habits, transport methods, food, goods, etc.) should come first.

ACCOMMODATION

Choosing low-impact accommodation can help to reduce your overall footprint (as well as persuading other hotels to go green). Look for small-scale properties that use renewable energy, have strict energy-saving measures in place and have a local sourcing policy. According to Responsible Travel, four-star hotels can have up to four times the carbon emissions of smaller, low-key properties. Rather than defaulting to large chain hotels, consider independent, family-run properties like those listed on iescape, White Line Hotels, Ecobnb and bookdifferent.

Seek out repurposed or renovated properties rather than new-builds. Buildings are responsible for approximately 20 to 30 per cent of the global carbon footprint. While hotels are a relatively small proportion of this, their resource consumption tends to be much higher than other buildings – they have 24-hour operations, guests are often reckless with energy and water use, and they have to support a massive variety of functions. In 2017, the International Tourism Partnership, working with the world's leading hotel chains, found that to curb climate change, the hotel sector would need to reduce its absolute emissions by 90 per cent. Meanwhile, most of the world's largest hotel chains are building new properties every day.

FOOD AND DRINK CHOICES

According to the University of Oxford, food is responsible for a quarter of global carbon emissions. Within that, 58 per cent of emissions come from animal products and 50 per cent of those come from beef and lamb. The IPCC says that switching to a plant-based diet can help address climate change; cutting meat and dairy can reduce your carbon footprint from food by two-thirds. Other experts recommend a 'planet-friendly' diet of reduced meat, reduced dairy, and an emphasis on local, seasonal and organic food.

Either way, knowing where your food comes from and how it is produced is fundamental to being a more eco traveller. For example, beef cattle raised on deforested land is responsible for twelve times more greenhouse gas emissions than cattle reared on natural pastures.

Responsible Travel's 2020 Holiday Carbon Footprint Report compared the foo(d)print from two holidays – one serving limited choice vegan meals and the other serving a more comprehensive mixed menu. Although not directly comparable, the latter's daily per guest foo(d)print is 11 kg CO_2-e, while the former's is only 3 kg CO_2-e.

'Your food is sometimes the single biggest, source of CO_2 emissions from your holiday. To get to net zero carbon by 2050, we'll need to fly less and change what we eat,' says Justin Francis, founder and CEO of Responsible Travel.

SUPPORT CHANGE

The travel industry is finally waking up to the climate crisis and some players are taking action.

One notable initiative is Tourism Declares A Climate Emergency, a collective of travel businesses that are developing climate plans in line with IPCC advice. The group was set up by adventure travel company Much Better Adventures and sustainable tourism expert Jeremy Smith. Founding signatories include some big names – Intrepid Travel, Exodus Travels and Explore – and they will work together to decarbonize travel.

Another way to champion progress is to support relevant campaigns. Responsible Travel's proposed Green Flying Duty advocates for 'a new global tax on aviation (based on a reformed version of the UK's APD) that will be ring-fenced for research and development in electric aviation and to improve railway connectivity'.

If you feel compelled to give up flying, join a local flight-free movement or pledge, like Flight Free UK or Flight Free USA.

Our biodiversity and climate crises are intricately linked. In tandem with reducing carbon emissions, we need to protect natural carbon sinks by supporting the conservation of forests, peat bogs, sea grass, reefs, mangroves, and a whole host of other ecosystems.

A note on offsetting: use caution

Everyone loves a quick win, which is one reason the carbon-offsetting industry has boomed. Offsetting involves calculating your carbon emissions and then buying the equivalent carbon 'credits' in order to 'cancel' them out. These credits, in theory, take the carbon out of the atmosphere by funding initiatives such as renewable tech or conservation work. You could offset emissions by providing clean cooking stoves in Malaysia, or planting trees in Scotland, or supporting elephant conservation through reforestation in Kenya.

It's by far the easiest way to address carbon emissions because it requires zero behavioural change. The offsetting industry is now worth around $500 million a year. It's a nice concept, but it is fraught with problems. Firstly, the industry is widely unregulated. A 2017 European Commission report discovered that 85 per cent of offsetting schemes don't follow through on what they promise. Carbon calculations and solutions vary wildly.

Secondly, offsets remove the incentive to make meaningful changes in behaviour. They appease guilt rather than solving underlying problems. This reliance on offsetting is a concern on an individual level, but it is even more worrying at a corporate one. For example, offsetting allows Heathrow to say that it plans to become 'carbon neutral' by 2030, even though it also plans to build a third runway that will be able to take an extra 265,000 flights a year.

Most sustainable travel experts recommend reducing your carbon footprint where possible and offsetting what's left. If you are offsetting, look for schemes with the Gold Standard label, or those that provide a verified emissions certificate. Ask how permanent the project is, whether it takes into consideration local needs, and whether it has been independently verified. Look for projects that protect existing forests, empower local communities, or invest in clean tech. The older the tree, the more carbon it absorbs, and so protecting our existing forests often has a greater impact than planting new trees. Schemes along these lines include Cool Earth, the World Land Trust, Forests Without Frontiers and TreeSisters. Atmosfair (see page 29) is widely considered to be the most robust carbon measurement and offsetting tool.

Planning a trip

Although it may seem like there's a lot to consider when planning a sustainable trip, it's worth remembering that taking a few actions is better than taking none at all – no single trip is going to be perfect and our travel choices sit within our broader impact on the planet. Someone might choose to be vegan in order to better justify a flight taken once a year, another might forgo flights altogether because they prefer to eat meat. We simply need to chip away at the various negative impacts we make and prioritize positive ones where possible.

Once you're armed with facts and practical information, it will become easier to rely on intuition and instinct to navigate what is and isn't a responsible and sustainable travel choice. And it will soon become clear that the most sustainable option is often the most inspiring one, too. This chapter will provide the tools you need to plan where to go, how to get there, where to stay and what to do.

Where to go

The world is more accessible than ever. Tomorrow, from my London base, I could nip down to the Mediterranean for little more than £50, or perhaps whizz across the pond to New York for just a bit more. Accommodation won't break the bank, either. I can rent a room on a site like Airbnb for £30 a night, or find a last-minute deal on any number of hotel booking sites. Admin and logistics are minimal – all I need is my phone, laptop and a Wi-Fi connection, and I'm off, free as a bird.

In many ways, this freedom is a great thing. Travel was once reserved for a wealthy elite, but now, thanks to low-cost airlines and cheap accommodation, it's become a more even playing field. Technology has played its part, too: work culture is more flexible and logistical barriers have been broken down by global connectivity. The world is smaller and it's ours for the taking, whether as a tourist, business traveller, free-spirited nomad or global citizen.

The problem is volume. The World Tourism Organization forecasts that international tourist arrivals will reach 1.8 billion in 2030. In Southeast Asia, a region that's seeing one of the world's most significant increases in tourism, destinations such as Vietnam are experiencing a 30 per cent rise in tourists every year. While this growth creates jobs, can help to protect wildlife and accounts for 10 per cent of the world's economy, it also comes at a price to the planet. What follows are some points to consider before you choose your next destination.

IVM·FECIT

The problem of overtourism

Overtourism refers to tourism's rising toll on the environment and local communities – the word was shortlisted by the Oxford English Dictionary as 'word of the year' in 2018. Destinations that previously welcomed visitors with open arms now think twice about whether the economic benefits outweigh the negative impact.

Take Barcelona as an example. The phrases 'tourists go home' and 'tourist: your luxury trip is my daily misery' can be seen spray-painted all over Barcelona – a bit of a downer for tourists to Spain's most visited city. Barcelona receives 32 million visitors a year, which is twenty times the resident population, and many locals have simply had enough.

In 2018, non-profit organization The Travel Foundation released a report outlining the 'invisible burden of tourism' and ways in which the industry can address it. Beyond over-crowding and the disruption of local lives and wildlife, the report highlights hidden challenges: pressure on services including sewage, energy and water, as well as higher rents and a shift in the type of amenities on offer.

Each year, 24 million visitors besiege Italy's already fragile floating city, Venice. Over half of them are day-trippers, barely contributing to the city's economy. The city was built for 150,000 residents, but today only 53,000 remain: rent is too expensive, pushed up by holiday rental schemes like Airbnb, and crowds render the city uninhabitable. Jane da Mosto, co-founder of the campaign We Are Venice, recently told CNN: 'The residents suffer very much, and it's beginning to feel like we're being deprived of our civil rights.'

In other places, there's been a more subtle transformation. In Dubrovnik, an influx of cruise passengers has fuelled the replacement of local amenities, such as barbers, restaurants, butchers

STOP AND THINK: BOYCOTTS

Some people argue that it's best to boycott countries that have poor human rights records or authoritarian regimes. Reason follows that by spending money in these places, you may end up funding human rights abuses, genocide or hate crimes. Examples include Myanmar, which itself called for an international tourism boycott when de facto leader Aung San Suu Kyi was put under house arrest. Another is Saudi Arabia, which is now opening up to tourists, but many journalists feel uncomfortable visiting given the murder of Jamal Khashoggi and the lack of rights for women.

The problem is, it's hard to know where to stop. Our world is so interconnected that few places have a clean slate. There's one school of thought that suggests that it is people living under difficult regimes or with conservative politics that are most in need of contact with the rest of the world, in order to provide a sign of hope, a source of information, or an opportunity to tell their story. So when it comes to boycotts, there's no easy answer.

and greengrocers, with rows of souvenir shops. And in Queensland, Australia, the food scene increasingly caters to wealthy visitors rather than locals.

It's not just a problem restricted to towns and cities. In 2018, mountaineer Nirmal 'Nims' Purja took a photo of a long queue on Everest's treacherous Hillary Step, which prompted questions about how many people should be allowed to climb the world's highest mountain at any one time. Meanwhile, over in Peru, a new airport proposal for UNESCO's Machu Picchu, which at that time was receiving up to 5,000 visitors a day, sparked worldwide petitions of protest.

In April 2018, the president of the Philippines closed the island nation's most famous resort, Boracay, calling it a 'smelly cesspool'. It reopened six months later with more than a new lick of paint; hotels that didn't comply with environmental regulation were pulled down, water sports were banned and casinos remain shut. Earlier that year, Thailand closed Maya Bay to allow coral reefs to regenerate.

Only weeks after the COVID-19 lockdown was eased, national parks struggled with overflowing toilets, epic landscapes were clogged with cars and snap-happy hoards trampled fragile ecosystems. From Iceland to Amsterdam, the Great Barrier Reef to Yellowstone, it seems that there is no destination insulated from the unprecedented rise in global tourism.

With so many examples of the negative impact of tourism, it's enough to make you want to pack it all in. But overtourism also presents us with an opportunity. It reminds us to think outside of the box, explore roads less travelled, listen to locals and create unique itineraries. If we embrace and respond to the challenge, we can reduce tourism's burden while improving our experiences.

Ways to tackle overtourism

With so many people travelling around the world, overtourism is a moving beast. One minute somewhere is unknown, the next it's flooded with visitors. While being mindful of this variability, here are a few ways to help tackle overtourism when choosing a destination.

SEEK LESSER-KNOWN DESTINATIONS

Ditch the bucket lists, the top-ten listicles and the Instagram hotspots, and start thinking imaginatively about where to go. If one of the key motivations to travel is to find new perspectives and ignite curiosity, you're far more likely to do that in a place that's less well-known – somewhere that you can discover and explore, rather than tick off a stream of recommendations. Who wants to queue for a view, hike with hundreds of others or be booed at by locals, anyway?

Each year, the World Tourism Organization ranks the least-visited countries in the world, and it includes such gems as the South Pacific island state of Tuvalu, the volcano-dominated Caribbean island of Montserrat, the palm-fringed beaches of Sierra Leone (above) and the UNESCO Biosphere Reserve on tiny island of Príncipe.

These places are less burdened by tourism and therefore can be more welcoming. Better still, in areas less familiar with regular visitors, residents can be indifferent to their presence, leading to perhaps the most authentic (for want of a better word) experiences of all. When places aren't geared towards tourism, there's less tourist-facing infrastructure to get between you and the local way of life. Not only will your experience be better, but your money is more likely to go into the hands of local business owners.

If heading to a remote island is not practical, there are simpler alternatives to consider. Eastern Europe is often overlooked, despite some incredible coastlines, mountain ranges and cobbled city centres that rival their western neighbours. The new Transcaucasian Trail linking hiking routes across Armenia, Georgia and Azerbaijan offers a far more intrepid and peaceful experience than walking Spain or Portugal's world-famous pilgrimage routes.

Colombia and Belize are fringed with the same azure waters as the Caribbean islands, and they come without the cruise day-trippers. Papua New Guinea (above) has a fraction of the visitors compared to its busy Southeast Asian neighbours. And, although politically unsettled in some areas, much of the Middle East welcomes visitors with open arms and near-deserted world heritage sites. If you're concerned about safety, tour operators are available to give advice, alongside your Foreign Office.

Capital cities often become tourist hubs, so try second cities and lesser-visited towns instead. Gothenburg will suit outdoor-lovers better than hectic Stockholm, Utrecht matches Amsterdam's architectural charm, but with an added dose of student grist, and Chicago's regenerating industrial pulse can offer a broader spectrum of American society than Washington DC.

Of course, alternatives are can mean a little compromise. I'm sure Lille is wonderful in its own right, but I doubt it's a substitute for Paris. And although I would rather do a tapas-crawl around Seville's cobbled streets than queue for Barcelona's La Sagrada Familia, I'm lucky enough to have done both.

You can still get off the beaten track in a well-known destination. Listen to local advice or seek out local guides. Rather than making a beeline for the UNESCO World Heritage site, ask locals if there's an alternative. Listings like Airbnb

NADA HOSKING, DIRECTOR, GLOBAL HERITAGE FUND

The Global Heritage Fund upskills local communities to transform cultural heritage into economic assets, enabling them to generate income and tackle depopulation.

–

What's next for sustainable travel?

Travellers are now looking for authentic experiences that foster connections to people. They want to learn about culture and history in a way that feels more meaningful than just checking items off a bucket list.

The communities living around travel destinations have the opportunity to directly benefit from this shift and protect their endangered heritage. In many parts of the world, historic sites and traditions are in danger of vanishing forever. Rather than protecting their cultural heritage sites and continuing their cultural traditions, younger generations are pressured to migrate to urban centres in search of employment. Sustainable travel is a powerful way to address community impact that is both social and economic and aims to ensure money spent on a tour or a trip stays in the community.

What gives you most hope for a more sustainable future?

It's heartening to see people finally addressing sustainability as a holistic concept. Sustainable travel is about taking comprehensive action. It's not just about reducing the number of plastic straws we use; it's about thinking of the larger, long-term impacts of travel and ensuring destinations are cared for and managed appropriately. I'm also encouraged by the new emphasis on slow travel. Of course, this depends on having resources, with time a key consideration, but I highly suggest slow travel for anybody wanting to forge deeper connections with communities.

What advice can you offer those looking to travel more sustainably?

Pick lesser-known destinations. This will help protect historic sites and provide a more enriching cultural experience, especially when you're not battling hordes of other visitors. The longer a traveller can spend in one area, the better. It benefits the destination because the money stays in that spot. And it's better for the traveller, because they get to know the real character of a destination. To really help, spending should stay in the local communities. In some places, it's made clear that money goes to local guides and accommodation. For other destinations, visitors need to do more research.

Experiences, Spotted by Locals and I Like Local can offer refreshing alternatives to staid sights and tours.

In the much-loved Balearic Islands (opposite, left), a recent sustainable tourism tax has helped to fund walking and horse-riding trails that crisscross the islands and navigate people away from busy beaches. In Majorca, paths are linked up with a series of mountain and hillside refuges so visitors can explore the entire island on foot. In Thompson Okanagan, Canada, the tourism association uses 'big data' to track and influence where people go – dispersing crowds with promotions and creating food and adventure activities to diversify what's on offer.

Some travel companies are rising to the challenge, too. Megan Devenish, Head of Product Expansion and Sustainability at Much Better Adventures, explains: 'We create trips that help you experience familiar places in unfamiliar ways.' Examples include finding less popular routes up Ben Nevis and Snowdon, as well as an alternative trek in Nepal that steers adventurers away from the overcrowded Annapurna circuit.

SUPPORT SUSTAINABLE PLACES

There are so many experiences waiting to be discovered: strolling along the Danube through the car-free heart of Vienna; nipping between cobbled lanes and parks in one of Ljubljana's free electric-powered Kavalir buggies; soaking up 6 per cent of world's biodiversity in one of Costa Rica's many protected areas; and hiking between Guyana's ecolodges. Wherever you choose to go, it doesn't take long to realize that sustainable places often make more interesting and exciting holiday destinations.

By travelling to cities and countries that support sustainable development, we're not only less likely to have a detrimental impact on people and the environment, but we're also proving that the investment made by governments, locals and all those involved has been worthwhile. We can also create a knock-on effect: by being more selective about where we go, we could help to create a ripple of positive change around the world.

The most responsible destinations put the needs of locals first. Happy locals mean better transport, renewable energy, accessible cities and the protection of green spaces and the environment. In 2019, the Global Destination Sustainability Index ranked Gothenburg, Copenhagen and Zurich (opposite, right) as showing the greatest improvement.

France, Switzerland and Denmark often rank highly for being the most liveable countries and having the best sustainable infrastructure. Costa Rica, Guyana, Vanuatu, Palau, and Chile have also recently been praised for a high-value, low-impact approach to tourism.

The West Indies island of Saint Kitts has a whole department working to ensure that tourism development is sustainable. Slogans 'Pro Planet, Pro People' and 'Good for Us, Better for All' shape the island's tourism strategy. Each year, the Saint Kitts Sustainable Destination Council runs a survey to gauge resident opinion: the latest revealed that 80 per cent of locals consider tourism to be a good thing.

If you're unsure whether somewhere has a responsible approach to tourism or not, look for places that promote green experiences and talk honestly about reducing visitor numbers. Look out for a tourist tax, as well as policies banning cruise ships or capping visitor numbers.

Bhutan's high entry price naturally limits numbers, while Botswana is often recognized for its conservation-led tourism. Amsterdam, well aware of its overtourism problem, has launched an 'Untourist Guide' to the city to encourage visitors to make a positive contribution.

For truly intrepid adventures, it's a good idea to head somewhere that doesn't have any marketing slogans or campaigns at all. Or, if you're comfortable taking the risk, you could even find inspiration on lists of where *not* to go. Either way, avoid tourism all together by visiting a place rather than a destination.

STOP AND THINK: INSTAGRAM

Horseshoe Bend is a dramatic U-shaped meander in the Colorado River, through the quiet, sandstone landscape of Arizona. For years, the view had been enjoyed mainly by locals. But two years after Instagram launched, visitor numbers rose from a few thousand before 2010 to over one million in 2018. The sudden surge took wardens and locals by surprise and soon facilities couldn't cope, the dirt track leading to the viewpoint became impassable and railings were put up after someone fell to their death.

It's not an isolated case. From Iceland, where a canyon viewpoint closed after hordes of visitors flocked to stand in the same spot as Justin Bieber, to Mumbai, where selfies have been banned after several tourist deaths, countries are experiencing a spike in travellers. While a guidebook might inform thousands of people over several years about a particularly idyllic or photogenic place, Instagram can do the same in seconds. Through geotagging – marking exactly where a photo has been taken – people don't even need a map to head to the same spot.

If in doubt, don't geotag. And, if you're after an authentic adventure, maybe leave the smartphone at home.

EMBRACE HOME TURF

There's a misconception that the further you go, the better the travel experience gets. But in fact, some of my most memorable adventures have been on home turf. There was the unexpected fight for survival after getting stranded on Scotland's tiny Isle of Eigg with no supplies or open shops (opposite, above), as well as the time I faced a blizzard at the top of Scafell Pike, England's highest mountain (opposite, below). I've been stuck in a bog while wild camping in Scotland, swam with giant barrel jellyfish off the coast of Dartmouth, Devon, and while enjoying a moonlit walk across the South Downs, I was chased by horses – an unexpectedly welcome adrenalin rush!

Staycations are holidays in your own country – whether exploring its far-flung corners or staying closer to home. They're not only better for the environment, cutting the carbon emissions of travel considerably, but they can make us more content, better citizens and ambassadors of wildlife and culture on our doorstep. All these things require time and commitment, but if we're too quick to jet off and explore foreign lands, we may be compromising our relationship with home.

Rampant advertising makes us long for more exotic climes. Advertising and social media can make jetting off as tempting as fast fashion. There's the envy factor, too – if they can have that, why can't I?

In 2009, philosopher Alain de Botton launched 'The Holiday at Home Box', including instructions such as: 'Try flying First Class. Get an ordinary armchair. Leave half a metre between the edge of the chair and a wall. Position a television close by. Imagine that you have paid in the region of £5,000 to be sitting in the seat.' He's got a point. When you think about it, the 6 a.m. start, the airport security, the packing for all eventualities and the pressure of spending a considerable chunk of your cash on one trip isn't always what you're looking for from a break. Having been blasted with so many idyllic images before even getting somewhere, the reality doesn't always live up to the hype.

A staycation, on the other hand, leaves room for spontaneity, surprise and delight. It can also be more relaxing; by removing the pressure to tick off activities and sights you *must* see, it's easier just to be in the moment. Whatever this type of travel brings, rain or shine, disaster or triumph, it's almost always the most sustainable choice.

Staying closer to home doesn't mean missing out; wherever you are in the world, adventure is closer than you think. In Canada, Parkbus links cities with nature by providing an affordable and accessible transit from the high-rises of Toronto to the beauty of Algonquin Provincial Park or the Bruce Peninsula. In Colombia, a cable car connects smog-filled Medellín with 54 miles of walkable trails amongst 38,000 trees, and in Wales, a disused quarry is now home to the world's longest manmade surf wave.

CÉLINE COUSTEAU, FILM-MAKER AND CONSERVATIONIST

Granddaughter of underwater explorer, Jacques Cousteau, Céline is a film-maker raising awareness of conservation and ethical issues and an ambassador for the Treadright Foundation.

–

What's next for sustainable travel?

I think more people are reevaluating what sustainable travel means. It's not just about taking bamboo cutlery with you, but readdressing why and how we travel, and what the purpose is. It's increasingly seen as an opportunity to give back and become further educated. People want to feel good about their choices and experience something meaningful. If that's not on people's minds then it's our job to make them realize travel can shift consciousness. Transformative travel is going to be the next big thing – it's about self, it's about others and it's about deepening the experience.

What gives you most hope for a more sustainable future?

That more and more people are having the conversation around sustainability. My hope is that people actually implement the more positive actions. It's now the responsibility of travel companies and service providers to step up to the challenge. Sustainability should no longer be the add-on or the exception; it should be the norm.

What advice can you offer those looking to travel more sustainably?

Evaluate your reasons for travelling and your destinations to find out how you can have the most positive impact. Take responsibility for yourself and consider what actions can you take, including your behaviour towards people and animals, taking reusables, and choosing what you eat. Scrutinize who you travel with – do they have sustainability practices within the company and are they giving back to destinations? We have to demand more action and check that proper sustainability measures are in place and adhered to. And if you can, donate to a cause that means something to you in the destination you visit. Even small amounts can make all the difference. Always do thorough back-checks and talk to operators before donating to ensure money goes into the right hands.

What to spend

It's widely believed that sustainable travel involves paying more money for a less luxurious experience. In fact, this couldn't be further from the truth. While it's true that implementing green technology and sustainable practices requires upfront investment, and this cost can trickle down to the guest, on the whole, where a sustainable travel experience is more expensive, it's often because it's superior.

Rather than paying for fine dining and chandeliers, you'll be paying for exclusive access to an untamed wilderness, an untapped culture, or guides that are experts in conservation or local heritage. Staff might be better paid, and therefore more likely to provide better service, or buildings might be constructed from more expensive local materials, but therefore be more inspiring and memorable to look at and be in.

Sustainable travel can also be the most budget-friendly way to get around. Rather than paying a premium for well-known chain hotels or restaurants, opt for the local taverna, or a street food market. Wild camping, bothies and mountain refuges are by far the greenest places to stay, and mostly free. Public transport is more affordable than jumping in taxis almost everywhere, and walking doesn't cost a dime.

When to go

Most destinations experience a spike in visitors at certain times of the year, and this can cause economic and overcrowding issues. Destinations that rely too heavily on peak seasons, therefore, drive locals away, and end up like ghost towns for half of the year – this vicious cycle can end up destroying a place and turning it into a theme park.

Travelling off-season (sometimes known as the green season) is a good way to spread the load. It's also more enjoyable to experience a place when the locals are more relaxed. It can be cheaper and easier to get around, as well as less of a logistical headache. In popular destinations, avoiding peak times can be equally good for locals; places suffering from overtourism need thoughtful travellers as much as anywhere else.

Of course, there are some downsides. Weather can be the primary driver of peak seasons, and we all love (and sometimes need) a bit of sun, although worryingly, with a more unpredictable climate this is becoming less of a guarantee. Some places can also shut down in the low season, with institutes such as museums closing for renovations, restaurants shutting up and services becoming limited. It's worth doing some background research before you book.

How to get there

The most sustainable way to get from A to B (as explored in 'Understanding carbon', pages 18–31) is often the slowest. This means ditching air travel and embarking on a journey by train, bus or ferry, or better yet, sailing, cycling or walking.

Unfortunately, when the call of travel beckons, we don't all have time to hit the road for months on end in order to go as slow as possible. With limited annual leave, jumping on a plane allows us maximum time away, with minimum travel time.

But things are changing. New rail routes are opening up, tour operators are offering flight-free packages, and schemes like Climate Perks are even providing extra holiday leave for those that decide not to fly, so slow travel is becoming easier. If you do fly somewhere, remember it's worth reducing carbon emissions once you get there by switching internal flights or taxis with slower ways and means.

Once you give it a go, you may never look back. Going slow is not only more environmentally friendly, it can also help us rediscover the joys of travel. Just as slow food celebrates the buying and preparing of food as much as the meal itself, slow travel celebrates the journey. Getting from A to B becomes an experience in itself, as you have time to soak up the minutiae of the places you travel through and better engage with the people you encounter. What follows are a few ideas to enable you to go slow.

Trains

The rhythmic clatter of a train on its tracks and the steady heaving of a locomotive across countries and continents is a wonderfully mindful antidote to our otherwise fast-paced lives. Whether hanging out of a train window in India or huddling over a bottle of vodka while travelling through Siberia, there's no better way to witness the gradual unfurling of geography and culture than on a long-distance train.

Rail tracks crisscross the world, and although some places are better suited to train travel than others, every continent has a great rail journey.

The beauty of trains is that there's something for everyone, whether that's the luxury of the Orient Express or the experience of sitting upright for days on the Trans-Mongolian Express.

Australia's The Ghan line cuts through the heart of the country, and the Reunification Express is a gentle way to experience Vietnam. Travelling by train is a quintessential experience in India, and with more than 7,000 train stations, there are no limits to where you can go. Harry Potter helped bring fame to Scotland's magical West Highland Line, and the Rupert Rocket zooms through British Columbia's unvisited nooks and crannies. Although the scenery is hard to catch whizzing through Japan on a Shinkansen, it's a fundamental part of Japanese culture, and no experience in China is quite as eye-opening as a rice wine-fuelled night on a third class cross-country train.

South America is well served by a coach network, but there are also a few standout rail journeys. Bolivia's Expreso del Sur meanders 300 km between the city of Oruro and the salt flats of Uyuni over seven hours, and the more luxurious Tren Crucero clatters through Ecuador from the Andes to the Pacific Ocean. In the United States, train lines are similarly sparse, but Amtrak have a few brilliant routes: look out for glass-top scenery carriages and running commentaries over the PA

system. To appreciate the scale of the country, head across the Rocky Mountains on the California Zephyr, or cruise the West Coast on the Coast Starlight from Los Angeles to Seattle.

Europe is raising its rail game, especially when it comes to sleeper trains: the Caledonian Sleeper from London to Scotland recently received a much-needed refurb; a new sleeper from Brussels to Vienna has made Central Europe more accessible; and Swedes can now hop across to London on an overnight train from Gothenburg. If you're after something a bit more lively, a music- and booze-fuelled 24-hour Jazz Train now runs between Amsterdam and Berlin several times a year. Meanwhile, Eurostar, one of the world's busiest international train crossings from London to Paris, Brussels and Amsterdam, is upping its eco game by banning plastic and committing to an overall reduction in carbon (despite already emitting only 6 g of CO_2 per passenger, compared with 133 g per passenger when flying).

Finding and booking rail routes is becoming easier too. For a definitive global guide, take a look at timetables and logistics on The Man in Seat 61, a website dedicated to rail travel. In Europe, Rail Europe has made finding a suitable itinerary and booking your tickets quicker than it's ever been, and Snow Carbon helps you book trains that will get you and your ski equipment to the slopes. Green Traveller has a series of flight-free guides including holiday-by-train ideas all over the UK and Europe.

Look for multi-day and rail passes where available to save money, and ask local tourist boards for advice. Tour operators including Original Travel, Contiki and Sunvil are featuring an increasing number of trips that forgo flights for trains. Meanwhile Planet Rail and Great Rail Journeys are raising the bar with 'platinum' trips. Long-time slow-travel advocates Inntravel use provincial trains to explore hidden pockets of a country, such as Spain's northern coast or Slovenia's less-visited lakes.

Ferries

For people-watching and seascape gazing, ferries are a top choice. They're also great for wildlife enthusiasts, as you might even spot dolphins and seabirds. If nothing else, docking at a port at your destination beats landing on a runway any day. Direct Ferries has comprehensive worldwide listings on its website, but here's a little taster.

On the south-west coast of Sweden, passenger ferries hop between car-free islands and quaint mainland villages that are dotted between Sweden's Gothenburg and Norway. The Cyclades, off the south coast of Greece, are best reached by the Blue Star Ferries network, and it's also possible to sail from Italy to the Greek island of Corfu by ferry. Combined 'rail and sail' tickets are also available in Scotland, Ireland and the Mediterranean, which make outlying islands cheaper and easier to access without flying.

North America is similarly well-served: Vancouver Island is best reached by ferry, and BC Ferries bob between Vancouver and British Columbia. Meanwhile, the Alaska Marine Highway System extends over 5,000 km of coastline, and the Steamship Authority in the United States runs seasonal services between some of New England's prettiest towns.

Several flotillas and ferries connect parts of Asia, too. Ferries link up South Korea with Japan and China, while Singapore, Indonesia, Malaysia, Thailand and the Philippines are all connected – these routes provide a great alternative to internal flights. It's also possible to cross the Mekong River in several places, from Thailand to Laos and from Cambodia to Vietnam.

Although still relatively carbon-intensive (it's one third of the impact of flying), ferries are slowly greening up. In 2019, the world's first electric ferry, E-ferry Ellen, made its maiden voyage connecting the island of Aerø to mainland Denmark. Stena Estrid, on the busy Holyhead to

Dublin route, is among the most energy efficient ferries in the world, and a number of battery-hybrid ships are in operation on routes provided by Caledonian MacBrayne Clyde & Hebridean Ferries in Scotland and on the Solent.

Sailing

Nothing feels quite as intrepid as heading out to sea as the captain of your own boat: being at the mercy of nature, following in the footsteps of explorers long gone, and snatching a glimpse of life lived out in the big blue.

Crewseekers is the best place to start for affordable sailing experiences. It's a little like Tinder for sailing crews: captains advertise for crew to work on yachts that are booked for around-the-world trips, coastal day sails and everything in between. It's a great way to test your sea legs or embark on the voyage of a lifetime with no cost to you but your time.

Once you've passed a recognized sailing qualification, like Day Skipper, it's possible to hire a sailing boat anywhere in the world. People advertise boats to rent on sites like Click & Boat, or Sunsail is better suited for holidays. If you prefer someone else to do the hard work, take a look at Happy Charter, which offers boats with a full crew, from traditional gulets to racing catamarans.

Expedition boats

Expedition sailing trips take guests to survey wildlife, monitor ocean plastic, or other 'citizen science' projects. It's a great and useful hands-on way to experience life at sea in safe hands.

Pangea Exploration's plastic and wildlife survey expeditions are some of the most sustainable on the market – its hulled sailing vessel, *Sea Dragon*, takes up to fifteen people out to sea from the United State's West Coast for between one to four weeks at a time. Research is meaningful (there's no tokenism here) and the crew is experienced, having travelled over 20,000 nautical miles.

In Europe, marine conservation group Wildsea organizes similar trips, including surveying dolphins off the coast of Italy and scuba-diving among seagrass in Ireland. St Hilda Sea Adventures take small groups to Scotland's outlying islands on converted lifeboats and fishing vessels, and Responsible Travel lists worldwide, eco-minded, small-scale expedition cruises.

Container ships

Landlubbers might prefer to jump on a container ship. These 100,000-tonne beasts take passengers thousands of miles around the world while they cart their cargo across oceans. Shipping routes dictate round-about voyages, making this the ultimate slow travel experience – it will take ten days to get from Vancouver to South Korea and fifty-five days from the UK to South America. As a paying guest, you're treated to a comfortable suite, hearty meals and have free-range of the ship including the bridge, deck and mess. From orcas and albatross to BBQs and karaoke, life on board is more varied than you might think.

Of course, container ships come with a carbon footprint, but by being one of a handful (or sometimes the only) passenger, you're hardly contributing to their raison d'être. Several agencies, like Strand Travel and Freighter Trips, will help to plan your journey and make bookings.

Rivers and canals

If your sea legs aren't ready for open waters, there are calmer ways to travel by boat. Rivers and canals don't just carve through geography and get up close to wildlife; these waterways also get to the heart of culture. Rivers were at the centre of trade and transport before roads came to rule.

That said, large-scale, irresponsible river cruising can do irrecoverable damage, so always ask questions to ensure passenger numbers are limited (no more than forty), waste (especially sewage) is responsibly disposed of, and the ship is efficient in its use of fuel. Itineraries should support rather than burden cities, villages and local people.

If you'd rather not join others, you can rent a canal or motorboat. Where possible, choose suppliers using renewable energy. We Are On A Boat rents out boats that use 100 per cent renewable-energy. In the UK, Anglo Welsh's newer boats use LED lights and more efficient fuel, and Waterways Holidays has a few electric vessels to hire. To embrace the sharing economy, the company Borrow a Boat helps customers do exactly that.

Road trips

Nothing but you, a cosy bed, camping stove and the open road: campervans are synonymous with freedom, adventure and the great outdoors. They fare well environmentally, too. Staying in your vehicle means slashing vast amounts of carbon associated with accommodation and hotel activities.

Although slow to market, electric campers are on their way. Spoot and eDub rent electric vehicles in the UK, and Camptoo allows you to rent motorhomes and caravans from private owners around the UK.

In New Zealand, Britz offers some of the most affordable electric RVs to rent, and the country's leading RV rental company, Tourism Holdings Limited (THL), has strong sustainability credentials. THL's 'Protect, Grow, Respect' policy includes carbon reduction targets, going plastic-free and electric vehicles. They have fitted telematics systems to their vehicles, which help to monitor fuel-efficient driving practices. The rental company also ensures that visitors give back to local communities via the Tiaki Promise, which encourages travellers throughout New Zealand to feel a sense of connection and responsibility toward the places they visit.

Taking to the open road isn't necessarily more carbon-efficient than flying. The impact of driving varies considerably depending on how many people are in a car, how new the model of the car is and what it runs on – diesel, petrol or electricity.

Simple actions like car sharing and switching the car for public transport, foot or bike where possible, all help to reduce kilometres on the road. Apps like Liftshare create opportunities to carpool, and park4night shows you available camper parking spots in towns and cities. A holiday can be the perfect time to trial an electric vehicle.

Cycling and walking

Whether stomping along ancient pilgrimage routes or whizzing along disused railways, far afield or close to home, travel doesn't get much simpler or more rewarding than relying on your own steam. With national trails, cycling tracks and quiet roads carving through landscapes, countries and continents all over the world, there are endless opportunities to walk or cycle. And they're almost always free.

Rather than battle crowds on the Camino de Santiago or Machu Picchu, battle with the elements on Patagonia's El Circuito, get up close to the jungle on Dominica's Waitukubuli National Trail or discover long-forgotten ruins on Turkey's Lycian Way. Cycle through central Europe on the Danube's car-free cycleways, or take on the Great Divide, a mountain bike route from Jasper, Canada to New Mexico.

Cycling and walking have substantial health benefits, too: mindful concentration, days filled with fresh air, a digital detox and the chance to get fit are just some of the positive side-effects. Contrary to popular belief, you don't need to be a fitness freak to plan a walking or cycling holiday. You can take it slow and watch your stamina and strength improve each day.

Look out for local organizations that bring people together to enjoy the great outdoors. The British Pilgrimage Trust invites walkers to 'bring their own beliefs' on walks around the UK, and cyclists offer to host fellow tourers all over the world via the website Warm Showers. Beyond that, there's plenty of help for travellers of all ages and preferences.

Inn Travel organizes self-guided walks between small, locally owned inns all over Europe, and Intrepid Travel makes use of public transport between worldwide hiking trips to reduce carbon emissions. Undiscovered Mountains specialize in trips in the Alps, and Adventure Alternative has spent twenty years compiling a worldwide network of responsible trekking companies.

Where possible, look for locally run hikes that give back to the communities they travel through. In Nepal, Sasane Sisterhood Trekking and Travel (partnered with G Adventures) trains survivors of human trafficking to become trekking guides in rural mountain villages, and Village Ways reinvigorates rural communities by organizing treks throughout the Himalayas.

For cycling enthusiasts, Skedaddle has recently launched flight-free trips from the UK, and Much Better Adventures has a series of self-powered adventures from Alaska to Costa Rica. H+I Adventures has teamed up with the Enduro World Series to create mountain biking trips on the Portuguese island of Madeira and in Finale Ligure in northern Italy. Sustainability is top of the agenda here, with a promise that 70 per cent of income remains in the country. KE Adventure explores Madagascar's Highlands by bike, and Slow Cyclist's tours aren't short on charm, with routes through Rwanda, Romania and more.

STOP AND THINK: CRUISE

Cruising is far from responsible travel, as we can see from the environmental fines that are piling up. In June 2019, Princess Cruises, and its parent company Carnival Corporation, were made to pay $20 million for dumping oily waste at sea. One cruise ship can produce up to 7 tonnes of waste a day, and many use notoriously dirty bunker fuel. Cruise ships can also be responsible for overtourism, unleashing thousands of visitors at a port at any one time. While some efforts are being made to clean-up the cruise industry – the world's first battery-hybrid cruise ship, Hurtigruten's MS *Roald Amundsen*, sailed the Northwest Passage in 2019 – arguably, big cruise ships shouldn't be in our waters at all. Look for small operators with strict environmental and social credentials.

Where to stay

There are countless certificates, schemes and awards that all aim to guide you towards the most sustainable places to stay. Some are more helpful than others, but with a few guiding principles in your back pocket, it's simple enough to spot the hotels-with-a-heart for yourself.

The most sustainable places to stay tend to be small and locally run. While mountain refuges, youth hostels, bothies and campsites all naturally fall into this category, going green doesn't have to mean sacrificing luxury. Some of the world's most exclusive lodges and properties take sustainability seriously. After all, these places often have the means to invest in cutting-edge green technology, wildlife and local communities.

That said, it might mean rethinking 'luxury'. Enjoying blueberries from Sweden or caviar from Russia in the Maldives isn't really sustainable – even if hotels claim that they are offsetting the carbon emissions. Sustainable luxury celebrates the jewels of a destination, rather than detracting from it.

What does an eco-hotel look like? Whether a slick converted townhouse in an up-and-coming part of town or a family-run farmstay on the edge of wilderness, an eco-hotel is more about an attitude than something you can see or hold. Firstly, it's a happy place: employees, managers and owners are looked after and proud of their work and the whole place exudes a sense of purpose, sincerity and passion.

The most sustainable places to stay also want to share their vision. Rather than preaching or virtue signalling (no one wants that on holiday), they might share their ethos via something as simple as a conversation, a welcome note, or an impromptu back-of-house tour. By the end of your stay, the chances are you'll be fully on board, and maybe even a little changed.

When you find these places, you're unlikely to ever look back. They inspire, surprise and delight, offering so much more than a bed for the night. These hotels, hostels, guesthouses or lodges go far beyond sustainability box-ticking – they're custodians of people and place.

Restoration before construction

Building a brand new hotel, however much it might be focused on green, energy-conscious design, is rarely the most sustainable option. Not only is there the carbon associated with transporting goods and construction, but the runoff, noise and pollution from building sites. Instead, look for places to stay that have renovated or converted existing buildings.

Take the example of Riad Anayela in Marrakech's medina. Here, more than 100 local craftsmen helped to restore the 300-year-old building using *tadelakt*, a labour-intensive waterproof plaster. The result of this slow approach is visible throughout the five-room guesthouse, from the worn tiles underfoot to an ancient poem hammered into silver on the riad's doors.

In Brazil's Reserva do Ibitipoca, businessman and conservationist Renato Machado used eighteenth-century building principles to restore the 300-year-old Fazenda do Engenho farmhouse into a boutique hotel. This not only better preserved the heritage of the building and the surrounding ecosystem, but meant employing local craftsman and reviving traditional skills.

An alternative option is to pick places that have used low-impact construction techniques that are sensitive to the environment, landscape and surrounding community. EcoCamp in Patagonia introduced the world to 'geodesic' domes (above). Raised platforms, skylights, solar panels, insulated walls and renewable materials all ensure that guests can sleep under Chilean skies knowing that they are having minimal impact on the otherworldly landscape.

Sensitive design

Kasbah du Toubkal in Morocco (above) slots so seamlessly into its elevated location that it looks as if it has always been there, a centuries-old Kasbah as majestic as the mountains behind it. In reality, a large part of it was built in the last twenty years. In South Africa's Sabi Sands Game Reserve, Earth Lodge has been built into and from the soil. Elephants and lions often roam on the landscaped roof and the hotel can barely be seen from just 50 metres away. Mindful aesthetics and structures are not only more in keeping with landscapes and local heritage, but can also break down barriers between tourists and locals.

A building or structure should consider what the local community wants to see, which tends to mean enhancing natural and cultural assets rather than suppressing them. In Greece's undulating wilderness of Zagori, the owners of Aristi Mountain Resort & Villas designed the hotel to mimic village houses. The entire property was built using local stone and wood, window sizes are restricted and pretty paths and gardens weave between rooms, so it feels less incongruous within the local landscape.

Eco-design can also maximize energy efficiency, reduce waste and support biodiversity. Low-impact and traditional building techniques are often the most efficient. In Indonesia, Nikoi Island's lodge and villas were built using the local *alang-alang* grass, which is best suited to the climate. In Chumbe Island in Tanzania, open-air villas are positioned so that a natural airflow removes the need for air-conditioning.

Swimming pools can employ green design, too. Tidal pools reduce freshwater use, while natural pools (rather than concrete) save energy, avoid the use of chemicals and support wildlife, not to mention being better for your skin. Austria is full of them, having coined the term *schwimmteiche* in the 1980s. The most eco-friendly, like Can Marti Agroturismo in Ibiza, use reed beds to filter and clean the water.

BEN PUNDOLE, FOUNDER, *A HOTEL LIFE*

Ben is the Editor-in-Chief of online travel magazine A Hotel Life *and founded the 'Stay Plastic Free' initiative to help hotels stop offering single-use plastics.*

–

What's next for sustainable travel?

There will soon be a generation and mindset of progressive travellers who will choose their accommodation at multiple price points based on the impact it has on themselves and the environment. So I think accreditation will be very important; we need a scale of just how responsible/sustainable each property is. Hotels and airlines will be scored on whether they are plant-based, zero-waste, plastic-free, using renewable energy, low-impact, and community programmes.

What gives you most hope for a more sustainable future?

There's a growing social expectation for companies to do better or they'll lose business. It's exciting to finally see companies electing real change!

What advice can you offer those looking to travel more sustainably?

Start with the question of how. Do you need to fly? What's the most efficient route? Then, think through the personal essentials: a reusable water bottle, a travel spork, etc.; ensure your sheets and towels don't get changed daily; bring your own toiletries; eat plant-based, local and seasonal food. It may seem odd at first, but everything will become second nature in no time.

Energy efficiency

Besides from camping or spending the night in an off-grid bothy, the most energy-efficient places to stay tend to be small-scale.

Look for properties with renewable energy, energy reduction targets and energy efficiency, which may mean in-room green temperature controls, building insulation and LED lighting among other elements.

Solar panels are a no-brainer in some parts of the world. Finolhu Villas in the Maldives uses 100 per cent solar energy, and Bucuti & Tara Beach Resort in the Caribbean has installed more than 600 solar panels. Within a year, the resort reduced the external energy demand by 15 per cent and is now close to using 100 per cent renewable energy. Other places might rely on a combination of sustainable options – Wheatland Farm in the UK uses 100 per cent renewable electricity through a wind turbine, a solar hot water system and a backup supply provided by Good Energy.

In countries such as Sweden, Scotland and Costa Rica, where the national grid is heavily dependent on renewables, a property having its own energy supply is less critical than in destinations dependent on coal. Also be mindful that 'off-grid' might mean relying on a generator, which still uses a large amount of energy, and that 'carbon neutral' hotels may be achieving that via offsetting (see page 31).

Lodges and hotels can become hotbeds of sustainable innovation. The Brando, a luxury resort in French Polynesia (opposite), installed the world's first SWAC – Sea Water Air Conditioning – that uses cold water from deep in the ocean to cool air circulated throughout the hotel. And Whatley Manor Hotel & Spa in the UK is due to trial one of the world's first waste-to-energy machines.

Energy reduction measurements and targets should be more sophisticated than simply saying 'we try to reduce energy'. Family-run Carbis Bay

Hotel & Estate in the UK has installed an on-site 'energy centre' that uses the heat from electricity production to provide a constant flow of hot water to the property, removing the need for twelve gas-guzzling boilers.

Eliminating waste

It's estimated that 5.25 trillion pieces of plastic are floating in the open ocean and up to 100,000 marine animals are killed by plastic every year. Whether it's because we've waded through it or seen wildlife suffering on David Attenborough's *One Planet*, most of us know the planet has a plastic problem.

Hotels and other places to stay generate double the amount of waste as residents. After experiencing what she describes as a 'Styrofoam hotel breakfast' one morning, Travel Without Plastic founder Jo Hendrix worked out that 'a pretty average-sized hotel was contributing almost 400,000 items of plastic waste to landfill every year from breakfast cutlery and plates and bowls alone'.

While we need to change our behaviour as travellers, it's more imperative that hotels (along with tour operators and destinations) step up – and it's going to take more than banning plastic straws and replacing mini amenities with refillable bottles. Look for hotels that are reducing plastic both front- and back-of-house, as well as across the entire supply chain.

The Cayuga Collection's laidback lodges and intimate hotels throughout Central America work with suppliers to help reduce their waste. The knock-on effect is enormous, with beer providers introducing recycling schemes and veg sellers banning plastic. A little further south in Colombia, Blue Apple is an eco-minded beach club on Isla de Tierra Bomba. After seeing the impact of the hotel's waste, the founder, Portia Heart, established the Colombian Coast's first glass recycling plant. To maximize its impact, an electric trolley whizzes around the city collecting glass from bars and hotels, too.

Of course, waste isn't limited to plastic and glass. A staggering one third of the world's food production goes to waste. Most of that ends up in landfill, which releases methane into the atmosphere. According to the World Wildlife Fund, food waste contributes to 8 to 10 per cent of global greenhouse gas emissions.

Food waste culprits include all-you-can-eat buffets, extensive menus and out-of-season offerings. To cut back, places to stay need to measure and reduce food waste through portion control, local and seasonal menus and, perhaps, by being bold enough to put an end to the beloved breakfast buffet.

Most of the big hotel companies have installed technology to track food waste, which can save thousands of pounds a year. Marriott, for example, has committed to cut food waste by 50 per cent by 2025, and Accor has launched a 'Too Good to Waste' campaign. Dukes in London has created a new 'Respect Everything, Waste Nothing' tasting menu, and Summer Lodge Country House Hotel in Dorset offers food-waste cookery classes.

More impressive still are the places implementing a closed-loop (creating no waste) food system by serving food that they've grown, composting waste and creating energy from scraps in a biodigester. Inkaterra in Peru converts biodegradable matter into a biochar fertilizer, Lapa Rios Rainforest Lodge in Costa Rica invites guests to feed food waste to the pigs, and The Stratford in London partners with Pale Green Dot, which connects chefs with vegetable growers in

order to exchange composted food waste for fresh fruit and veg.

Saving water

Demand for water is expected to exceed supply by 40 per cent by 2030. By the same year, half the world's population will be living in areas of high water stress, while approximately 1 billion people in the world currently don't have access to safe drinking water.

Sustainable hotels and places to stay are mindful of water scarcity, putting local and environmental needs first. It's worth considering whether watering that golf course might be draining the local supply or if it's appropriate to build a new hotel with seven pools in a place where there's barely enough drinking water for local communities.

Nowhere understands this more acutely than Cape Town, where the buildup to 'zero water day', when the city ran out of water reserves, almost brought the city to a standstill in 2018. For several years beforehand, green pioneers like the Vineyard Hotel were raising awareness through the use of shower timers and by replacing bath plugs with rubber ducks to make people think twice before having a bath (you can go to reception to swap your duck for a bath plug if you want one). In Mumbai, The Orchid has installed a sewage treatment plant on its roof to use wastewater for the hotel gardens, and in

Thailand, beach resort Soneva Kiri has created a water reservoir and deep well to provide water in the dry season.

Other water-stressed countries include Qatar, Lebanon, Botswana and India. In places like Bali, the west coast of the United States and central Australia, there are rising concerns over limited water supplies, too. If hotels and lodges don't have water-saving measures in place, they are probably burying their head in the sand along with other environmental issues. Simple steps can include collecting rainwater, planting native species (that restore soil moisture and reduce the need for watering) and greywater recycling systems.

Local food and drink

As explored in 'Understanding carbon' (pages 18–31), food production and shipping food around the world, from fine wine to avocados, is a significant contributor to a holiday carbon footprint. Supporting produce from intensive farming adds another layer to the problem – pesticide runoff, water shortages, land degradation, poor labour rights and poor animal welfare, not to mention that this sort of farming does little to contribute to a local economy, are just some of the issues. So the 'get what you want anywhere in the world' food system is no longer fit for purpose.

The most sustainable places to stay grow produce (ideally look for no-dig or permaculture principles that are more sensitive to the soil and wildlife), limit food miles and are either vegan, vegetarian or have a strict organic, nose-to-tail and wildlife friendly meat policy. Fish should be approved by the Marine Stewardship Council, eggs and chickens organic and free-range, and products should be Fairtrade- and Soil Association-approved where possible.

Obviously, you're unlikely to trawl the labels of every piece of food you eat, but it's worth knowing that sustainable sourcing is a serious business. When asked, most responsibly minded chefs or restaurant owners are incredibly passionate about it.

Saorsa 1875 is the UK's first vegan hotel, serving everything from watermelon sashimi to reimagined haggis, and both Perché dans Le Perche and Country Lodge in northern France supply guests with hampers filled with produce from the region's burgeoning organic farms and producers. In Chile, guest lodge &Beyond Vira Vira ensures it is supporting local producers by purchasing supplies like milk, flour and nursery plants, but makes or grows its own cheese, bread and vegetables to cut back on plastic.

Recognizing the limitations of space, the small island resorts of Nikoi and Cempedak in Indonesia have developed a 7-hectare community permaculture farm on nearby Bintan. It's a win-win: locals have access to new jobs and get trained in permaculture principles, and an impressive 90 per cent of the resorts' food and drink is sourced locally or via the farm.

When it comes to drinks, the same rule applies: go local where possible. Opt for homemade syrups or local juices rather than default global brands, and embrace the rise of biodynamic and organic wines and small-scale brewing. The hotel SALT of Palmar in Mauritius introduces guests to locals that invite them in for homemade rum, while independent vineyard booking site Wine Cellar Door guides clients to vineyard stays in the UK. Look out for closed-loop innovations like UK-based Toast Ale that makes beer from the leftover bread that's collected from hotels, restaurants and cafés.

Nurturing people and communities

The concept of hospitality comes from the ancient Greek *xenia* – the generosity and courtesy shown to those who are far from home. But being a sustainable place to stay means caring for those on your doorstep, too. Tourism supports one in ten jobs around the world; with this economic benefit comes a responsibility to ensure work and income goes to those that need it most.

Responsible recruitment involves hiring more than 70 per cent of employees from the local area or working with social enterprises and other employment schemes to provide work to disadvantaged groups.

In Sri Lanka, Jungle Beach by Uga Escapes runs the East Coast Women's Empowerment Project to train local war widows to operate the laundry room. In East London, Good Hotel prides itself on training local unemployed people (above). Saira Hospitality is a global scheme that creates pop-up hotel schools for locals. In Germany, bio-hotel Landgut Stober was one of the first companies in Brandenburg to offer training and employment to refugees.

The Sumba Hospitality Foundation on the Indonesian island of Sumba goes one step further and exists to uplift and train the local community. Its Maringi Eco Resort is not only the island's most sustainable place to stay, but is also a training base for locals ensuring that they are the main benefactors of tourism.

Other initiatives include the Hotel School, a joint initiative between The Goring Hotel and homeless charity Passage in London, and Youth Career Initiative, which operates worldwide to help disadvantaged young people turn their lives around through apprenticeships in catering or hotels. These employees are also often more committed and loyal; responsible recruitment leads to better service.

Looking after employees is an important part of being a sustainable place to stay. It means paying a living wage, avoiding agency staff (who can be more likely to get entangled in labour rights abuse or modern slavery), and having strict policies in place for human rights, modern slavery and human trafficking. Staff well-being mustn't be overlooked, and that encompasses the provision of training and good living quarters, as well as exercise opportunities and a healthy canteen.

Community outreach doesn't stop at employment. In Yorkshire, UK, The Traddock uses funds earned from recycling to help keep the community swimming pool open. In Scotland, eco-campsite and farmstead Comrie Croft organizes an annual mountain biking festival for the surrounding villages, and on Chumbe Island in Tanzania an environmental education programme welcomes more than 300 students each year to understand coral reef conservation and sustainable fishing.

US-based chain Ace Hotels was one of the pioneers of community outreach. When launching in Chicago, Ace donated money from bookings to a cultural arts centre and youth literacy group in the city, and it also improves pay and conditions for musicians and sometimes adds a voluntary community levy on room rates.

Promoting biodiversity

Protecting biodiversity is everyone's responsibility, and accommodation providers, whether they own a towering skyscraper or an ecolodge, are well-placed to do their bit.

In the UK, the Knepp Castle Estate not only takes visitors on 'safaris' to better understand native flora and fauna, but also advocates for rewilding across the country. Its owners have been influential in a movement that's seen forest cover in the UK increase from 5 to 15 per cent.

STOP AND THINK: AIRBNB

A 2019 report by the Economic Policy Institute says that 'evidence suggests that the presence of Airbnb raises local housing costs' in American cities. Residents and officials all over Europe have also blamed the holiday rental site for pushing up rents and limiting housing stock, not to mention completely altering the fabric of a city. Although Airbnb was launched with good intentions – the sharing economy is a wonderfully accessible and sustainable alternative to corporate conglomerations – as with many other democratizing plans to come out of Silicon Valley, it's got some serious problems. A locally run guesthouse, hostel, hotel or bed and breakfast will be regulated, whereas Airbnb can slip under the radar. Alternatives include a house-swapping company like Love Home Swap, or Fairbnb – a non-profit with restrictions in place to prevent damage to local lives and housing. If you do use Airbnb, only choose properties that are someone's actual home – you can usually tell by the volume of bookable dates on offer.

In Wales, Old Lands has carefully converted self-catering barns to host the Gwent Wildlife Trust and holidaymakers. Thanks to their wildlife-friendly farming, the 80-acre estate logged its 3,000th species (they're counting plants, animals and fungi) in 2019. Hotel gardens can also support wildlife. In the garden of the Schwarzwald Panorama in Germany's Black Forest, beehives, insect hotels and birdboxes encourage wildlife to make the most of the hotel's wildflower meadow.

In the United States, Badger Creek Ranch in Colorado is leading local efforts to turn towards regenerative cattle farming, and Six Senses Laamu in The Maldives is protecting hectares of sea grass (which other resorts deliberately remove) – a vital carbon sink and habitat for endangered species like green turtles.

In a city, look for a hotel that greens up whatever space it has or works with schemes run by charities like Trees for Cities, which aim to improve air quality and biodiversity. For example, Thon Hotel EU, in the middle of busy Brussels, supports insects and bees with a green roof made of grasses and lavender, and the hotel Parkroyal Collection Pickering in Singapore features curved high-rise gardens, waterfalls and green walls.

How to identify a green hotel

Sustainable travel is so on-trend that it can be hard to identify those who don't just talk the talk, but actually walk the walk. With the above information, you should know what you're looking for – places with energy and waste targets, strict local food, material and employment sourcing, and a visible commitment to heritage and the natural environment. If a hotel or lodge website isn't forthcoming with this information, then ask as many questions as you need to feel satisfied. If they still aren't forthcoming, this probably means they're not

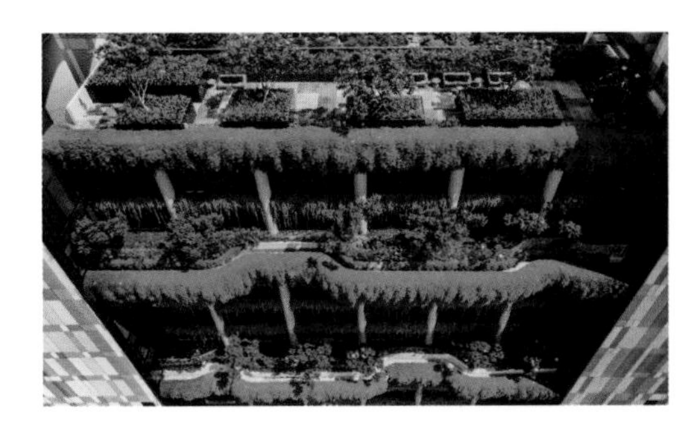

taking sustainability seriously. Sometimes you just have to trust your instincts!

Besides from relying on a property's website, search for press coverage: sustainability is a hot topic, so if somewhere is exemplary, the chances are it's been picked up by the press. It's also worth looking for a 360-degree approach to communicating sustainable targets and achievements – is it on their Instagram and Facebook pages, have they won awards, do they talk about sustainability when providing information on all experiences and rooms in the hotel? The most advanced green places to stay tend to proudly communicate how their sustainable approach enhances their offering.

Another indication of good intent is a reputable certification or label. Some certification schemes are more marketing platforms than a benchmark of serious sustainability credentials, so it's worth doing a bit of research. There are more than 200 out there, so it's impossible to summarize the pros and cons of each one. The most reliable are those that involve on-site assessments or are GSTC (Global Sustainable Tourism Council) approved.

If you're using an eco-hotel aggregator like Bookdifferent, Kiwano, Grean Pearls or Ecobnb, check that their criteria addresses energy, waste and water, as well as supporting the local community and ecosystem, and being a responsible employer. Avoid mainstream hotel aggregators that can take a large cut of profits from small, independent businesses, and instead book direct – you're likely to get a warmer welcome, too.

RANDY DURBAND, CHIEF EXECUTIVE OFFICER, GLOBAL SUSTAINABLE TOURISM COUNCIL

The Global Sustainable Tourism Council is an accreditation body for travel and tourism sustainable certifications, which are given based on merit, neutrality and compliance with international standards.

–

What's next for sustainable travel?

What must be next – and I think will be – is a broader awareness that all forms of travel must be made sustainable. We should erase from our minds images of eco-treehouses that virtually nobody visits. Sustainability must be mainstreamed. We travel mostly to cities, not places with treehouses. We need to act sustainably in all we do, including our normal and business travel. The planet is gravely in danger in many ways and we need to act accordingly. Immediately.

What gives you most hope for a more sustainable future?

I speak all over the world and, based on interactions across many cultures, two things give me hope. Firstly, everywhere I visit anyone under the age of thirty-five is motivated to act. Secondly, during the past year and a half we have seen a shift across all age groups and all cultures away from mostly talk towards a higher amount of action. Finally.

What advice can you offer those looking to travel more sustainably?

Don't feel too ashamed to fly, but fly rarely and only when needed, choosing cleaner energy alternatives when necessary. Also, help us increase from only 1 per cent of travellers choosing carefully versus the 72 per cent who say they intend to. Act!

What to pack

The three Rs are a useful guide when you shop and pack for an expedition, adventure or holiday: Reduce consumption and packaging, Reuse items and packaging, and Recycle what you can before hitting the road. If you are going to buy something new, invest in durable and high-quality items that will last – most sustainable manufacturers offer some kind of guarantee or repair service.

If you're flying, pack as light as possible to minimize carbon emissions on take-off and landing. Also, think about whether some things can be purchased from the destination – this can help cut emissions, support local economies and quench any retail therapy thirst by encouraging you to buy stuff you need rather than pointless tat.

Sustainable fashion

The fashion industry is responsible for more carbon than the aviation and shipping industry combined – whatever you spend, your new holiday wardrobe really can cost the earth.

Brands with B-Corporation status (one of the world's most robust sustainable business certifications) are a good place to start for sustainable kit – try Patagonia, Finisterre and Athleta, and changemakers like Batoko are leading the way with recycled plastic swimwear. Website Positive Luxury has a curated list of sustainable brands to help guide shoppers.
Better still, go secondhand: borrow from friends or hit the thrift and charity shops. Apps like Vinted, Swancy and Swooped all let users swap or buy secondhand clothes. The travel specialist Contiki recently teamed up with the #LoveNotLandfill campaign to run a series of holiday wardrobe clothes swap events. The first event saved 1,700 garments from landfill.

Plastic-free

Plastic is everyone's problem, and while it's easy to forget about that when you're tearing around packing for an adventure, it's more important than ever to be part of the solution when travelling. It's common to get through around 30 plastic bottles over a two-week holiday, and that's not to mention the straws, foam food containers, plastic bags and plastic cups you might use. Unbelievably, only 9 per cent of the world's plastic packaging is recycled.

There are several simple solutions. Pack a reusable water bottle and carry it everywhere. If you're travelling to a place where you can't drink the tap water, choose a water-filtering system like LifeStraw. Reusable bags can make a difference to your plastic waste, especially since not every country has regulations on plastic bag use. If you like a takeaway coffee, pack a reusable cup. When self-catering, camping or picnicking, take Tupperware, bamboo or reusable cutlery, and Bee Green Wraps (a sustainable answer to cling-film) can be handy.

A clear plastic washbag makes it possible to ditch the airport's plastic toiletry bags, while companies such as Lush and Friendly Soap have a range of shampoo and wash bars so you can forget miniatures. If you prefer to take your regular toiletries, buy refillable miniatures, that way you're not using and throwing away new plastic every time you travel. Or try out Circla – a new toiletry buying and returning scheme trialing in UK airports.

Plastic pollution in the form of microplastic – the hidden but deadly particles that are doing as much damage to ecosystems as the plastic we can see – is often used in sunscreens and other beauty products. Always use a sunscreen that states no microplastics and have been designed with the environment in mind, such as Waterlover Sun Milk by Biotherm or products by beauty brand Weleda.

For a purpose

It feels like the world of do-gooding has moved beyond philanthropic handouts. It's more sustainable to support those in need by funding local employment, products, social enterprises and services. However, when that's not possible, the charity Pack For A Purpose directs travellers to items that specific charities or communities need, which you may want to consider adding to your bag. Listings range from healthcare supplies for senior citizen homes in Cuba to school soccer kits needed in Nepal.

When you get there

Being a sustainable traveller doesn't stop at the planning stage. Every decision and activity you choose on holiday has an impact on the local community and ecosystem, so it's worth being mindful. Here are four things to have at the top of your mind.

Getting around

Whether sandwiched between chickens on a crowded Guatemalan bus, sweltering inside a local Caribbean ferry, or cruising between Berlin's parks on one of the many bike-sharing schemes, opting for local transport provides a deeper insight into a destination and reduces carbon emissions. Visitors often have access to different passes and transport cards, so check in with the local tourist board and do some planning before you travel.

Going local

Choose experiences that are run and led by locals. In Mexico City, Eat Like A Local (opposite) offers 'culinary safaris' that give 100 per cent back to local people, and in Thailand, tours with Bangkok Vanguard work against the tide of corporate development to ensure that tourism preserves rather than destroys the city's heart and soul. I Like Local, Tours By Local and Leap Local will each connect you with local guides. With Locals has a spread of (mostly food-related) activities and tours throughout Asia, while Embark connects travellers with local outdoor enthusiasts and guides for everything from bouldering to surfing.

If you're travelling with a tour operator, choose one that uses a local host. This means your money will stay in the destination, rather than going to a global company. It also leads to a more authentic experience. Check to see whether they commit to reducing waste, visit lesser-known sites, avoid overtourism, reduce carbon emissions where possible and contribute to community or conservation projects. If this information isn't available, don't be afraid to ask before booking. Responsible Travel has over 6,000 vetted trips, G Adventures gives each trip a 'Ripple Score' to show how much income stays in the destination, and Intrepid Travel and Better Places Travel are both certified B-Corporations (a highly regarded sustainability certification).

Respecting customs

Globalization and increasing tourist numbers have undermined customs and culture in some places. Ten years ago, tourists wouldn't have dreamed of walking through Marrakech's medina in skimpy clothes, but today these are more common than jellabas. While traditions should always be allowed to evolve (keeping people in a museum-state for tourists' sake is just as bad), it should be on local terms.

If in doubt, always respect customs regarding food, clothing and behaviour, even if others around you aren't. Doing so will often make you less conspicuous and threatening, and locals may be more likely to chat. Always ask permission before taking photos and follow local advice about cultural sensitivity.

For outdoor adventures, read up and obey the local code of conduct regarding wild fires, sacred sites, rubbish, human waste and protecting plants and wildlife.

Giving back

Travel can open your eyes to a standard of living far beneath your own, and many travellers feel compelled to give something back. This used to involve handouts – tour operators would actively encourage tourists to give money, pens or sweets to kids in need. Hopefully, we've waved goodbye to this form of do-gooding. Responsible travel does not rely on handouts, nor does it play into the hands of stereotypes or an 'us and them' mentality. It improves lives in the long-term through jobs, supporting enterprise initiatives, or funding projects created and chosen in close collaboration with the government or local leaders. If you want to give back, always do so through registered charities or by supporting social enterprises and locally run businesses.

CEM
11904

Regenerative travel

Sustainable travel is not one genre of travel, nor is it a set of criteria; it is a mindset that can be applied to every trip, motivation and destination. Whether raving all night in Berlin or trekking in the Himalayas, there are endless ways to increase the positive impact we're having on people and the planet.

Having a positive impact doesn't have to involve long-term volunteering. Instead, it may mean being more discerning about where we go, favouring destinations with strong environmental credentials, and places that put locals first. By choosing a specific tour, hotel or operator, we can show solidarity with a marginalized part of society, or champion places suffering from a natural or humanmade disaster. In cities, our choice of hotel might help to fund green innovations, or we could help to breakdown prejudices on a migrant tour. We can even use our travels to save a species from extinction.

Once we start to consider doing good every step of the way, a multiplier effect gets under way. By booking experiences, trips or accommodations that have a positive impact on local people and the environment, we're sending an important message to destinations and the travel industry.

We're not only talking about do-gooding, purpose and sustainability; we're choosing to spend our hard-earned cash on it. We're 'voting with our wallets' – letting the world know that we stand with the good guys and this is the future. Not only is this critical for people and the planet, but it will also lead to more fulfilling experiences. Here are a few ways we can travel to make the world a better a place.

Bolstering communities

A knowing smile that transcends language barriers, sharing a meal with strangers, or catching a glimpse of an indigenous ritual: it's meaningful interactions with people that often make travel experiences life-changing. And, we're hungrier than ever for them. In a world rife with fake news and systemic corruption, people on the ground are a vital source of hope and inspiration. Human stories are also an essential antidote to our increasingly digital lives.

There are plenty of experiences that connect travellers and communities. This could be through visiting a locally owned lodge or staying in a homestay. In South Africa, it could involve switching a game drive for a community visit, or in Peru, it might mean donating to a local charity or project that resonates. But is this enough?

According to the UN World Tourism Organization, only five dollars of every $100 spent in developing countries stay in that destination's economy. So even though we might think that by visiting somewhere we're supporting a local economy and therefore community, we may be lining someone else's pockets.

The most exemplary responsible travel experiences do more than merely foster connections; they make communities stronger, help people to become financially and socially independent, and empower the marginalized. Rather than focusing on what the traveller can get from interactions with the community, responsible experiences are based on mutual respect and a two-way dialogue.

Tourism can be an enormously powerful tool to uplift communities and breakdown stereotypes. Whether sharing skills or funding community-owned enterprises, seek out the changemakers transforming lives through travel. On the following pages are a few standout examples.

Sharing Skills

Learning from one another and exchanging skills is a great way to foster genuine connections. Skill-sharing also has a much more profound positive impact than token volunteering experiences (see below). Before embarking on any form of volunteering, start with your skillset – contemplate what your most valuable skill is, and explore where and how to share it with those that need it most. Look for specialist organizations that set up placements for specific professions, or turn to charities that advertise skill shortages. For example, in Nepal, there's a shortage of psychologists to deal with child-trauma and human-trafficking victims. To create a lasting impact and empower local people, look for placements or holidays that facilitate training or workshops. Go in with an open mind, and be prepared to learn just as much as you put in.

STOP AND THINK: VOLUNTOURISM

Building a school, feeding orphans, and digging a well are just a few of the activities that would have once been celebrated as responsible travel success stories. Today, however, voluntourism has a murkier reputation. Travellers could be taking away a much-needed source of income and employment from local people. Worse still, they could be illicit in corruption – where individuals profit from do-gooding tourists to the detriment of local lives. The most extreme example of this is orphanage tourism. It is estimated that of the 8 million or so children living in institutions worldwide, over 80 per cent of them have at least one living parent – many are coerced into orphanages to make money from tourists. Always scrutinize volunteering opportunities for ethics, check that providers have child-protection policies in place, and avoid orphanages.

CASE STUDY

ORBIS EXPEDITIONS, MALAWI

Malawi-based travel outfit Orbis Expeditions turns conventional voluntourism on its head. Rather than getting travellers to take part in activities that would be better suited to local craftsman and experts, its Skills Expeditions uses traveller knowledge to upskill local entrepreneurs. The highlight of a nine-day trip, which includes climbing Mount Mulanje, a tea plantation stay, and idyllic Lake Malawi island retreat, is an Entrepreneurs Forum where up to fourteen travellers share advice with Malawian entrepreneurs. Candidates on both sides are selected to ensure there's no tokenism.

Kate Webb, Orbis Expeditions founder, came up with the idea after living and working in Malawi and recognizing that there is barely any backing for small-to-medium businesses. She says, 'There's support for huge businesses or aid, but nothing in between. For these women, having an accountant, teacher, marketer or operations manager sit with them, share ideas and demonstrate how business works is life-changing.'

Two businesswomen to benefit from the experience is Trinitas, who produces reusable sanitary pads, and Endrina, who runs an organic vegetable and sunflower oil social enterprise. Both women empower others through their businesses. A lack of sanitary products in Malawi means that most girls don't attend school regularly, and by providing seeds on a cost-only basis, Endrina allows women to generate an income of their own via small-scale farming. After the last skill-sharing workshop, Endrina observed, 'I learned how to market products using social media; the importance of record-keeping; and the value of networking. We are now keen to pass this knowledge onto other businesses in Malawi.'

In Cartagena, not-for-profit Domino Volunteers has a similar ethos. Working with thirty-five social enterprises and charities around the colour-popping Colombian city, Domino takes a bottom-up approach; projects decide what help they need. Opportunities include everything from sports instructors to creatives and hair stylists. If you have accountancy skills under your belt, Accountancy for International Development (Afid) match accountants with charities that need financial support worldwide and Projects Abroad matches teachers with relevant training placements.

Supporting Enterprise

Moving on from a dependence on handouts or philanthropy, responsible tourism can best uplift communities in the long-term by creating economic opportunities. Rather than merely generating custom for shops, restaurants, and other services, travel companies need to support and facilitate enterprise – funding entrepreneurs, reaching out to communities with ideas, and guaranteeing a sustainable level of custom. Only when accommodations, destinations or tour operators take this active approach, does tourism genuinely improve local lives. As a traveller, support lodges and tour operators creating these meaningful projects and ask how you can help. Once at a destination, find out from the tourist board what locally owned enterprises you should support, whether visiting a particular market, going on a specific tour or buying particular souvenirs.

CASE STUDY

XPLORIO, SOUTH AFRICA

Another innovative concept that helps entrepreneurs around Walker Bay is Xplorio – an online services listing website designed to promote local businesses. By creating listings for smaller entrepreneurs, Xplorio brings them in line with mainstream business opportunities. The Xplorio directory expands the reach and discoverability of businesses that would otherwise get lost in Booking.com or Google ratings. Small, start-up cleaning businesses like Zwelihle have won long-term cleaning contracts thanks to being found on the Xplorio directory.

G-ADVENTURES AND PLANETERRA, WORLDWIDE

Global tour operator G-Adventures understands the importance of supporting local enterprise so acutely that in 2016 it developed a score to show the percentage of money it spends locally on each trip. It's called the Ripple Score – after the positive social 'ripple' effect each tour has. The score was created in partnership with Planeterra, a not-for-profit that works with G-Adventures and communities to create opportunities through tourism. Some key projects include Oodles of Noodles, helping former street kids in Vietnam find employment in restaurants, and Wiwa Tours, whereby the indigenous Wiwa community leads tourists on Colombia's Lost City trek.

GROOTBOS PRIVATE NATURE RESERVE, SOUTH AFRICA

In the heart of South Africa's Floral Kingdom, Grootbos Private Nature Reserve and its luxury lodges have had a staggering impact on local entrepreneurs. Perched above Walker Bay, with its sweeping sands and infamous whale sightings, the Grootbos Foundation ensures that the local communities, including the nearest townships, benefit from tourism. The foundation improves the lives of over 11,000 people by complementing state provision – elevating education, income, and health and well-being for everyone.

Siyakha is the foundation's employability and enterprise development programme. Schemes include organic farming, artisanal skills and eco-tourism initiatives. Some of the projects work directly with the lodge; Growing for the Future is an organic farm run by the community alongside a wax recycling candle-making initiative and hydroponics strawberry growing project. They all sell products back to Grootbos.

A Grootbos Foundation entrepreneurship programme has supported sixteen businesses with seed funding. It has also helped people like Siyabulela Blayi, the proud owner of Siya Fast Food, get back on track after his business burnt down due to unrest. Other successful business owners, like Olwethu, who owns an Internet café (vital for communications throughout the community), and Khuselwa who owns a takeaway business, have both received expansion funding from Siyakha.

Empowering People

Wherever you are in the world, from deep inside the Amazon to the heart of London, there will be marginalized groups of people struggling to make their voice and needs heard. Tourism is well-placed to help. In many destinations, travellers bring open minds eager to listen to and understand different perspectives. The tourism industry also tends to have a global mindset and is, therefore, more likely to challenge inequalities than other sectors. As travellers, we need to support accommodations or tour operators using tourism to empower marginalized people, whether through employment opportunities or providing a platform for lesser-heard voices and issues. Some of the most successful examples are those that address gender inequalities and uplift women, which has a positive knock-on effect because women are more likely to invest in education and community infrastructure. It's also good for climate change because it has a direct impact on overpopulation; educated women have fewer children, and ensure that their children are educated, too.

CASE STUDY

SUMBA, INDONESIA

The remote Indonesian island of Sumba welcomes fewer visitors than its neighbouring islands, but land speculators are keen to change that, coming in thick and fast in order to take advantage of Sumba's unspoiled waterfalls and wild beaches. Already more than 60 per cent of Sumba's beaches are privately owned; the land is often bought from locals at criminally low rates as investors seek to exploit the citizens of one of the region's poorest and most isolated communities.

While visiting the island in 2013, Belgian philanthropist Inge de Lathauwar desperately wanted to help make the situation fairer. Recognizing that she can't stop the tide of tourism, Inge has set her mind to empowering the local community to benefit from it. The result is Masungi Ecolodge, a lodge and hotel school designed to train Sumba locals in hospitality skills. By staying in one of the five bamboo pavilions or luxury villas, guests act as the guinea pigs for the trainees that are typically between seventeen and twenty-three years old. There are opportunities to support environmental initiatives too – helping out in the dedicated permaculture farm or learning about food waste, for example.

Alongside providing the skills needed to help locals gain employment in tourism, the school empowers them to have more of a say in their island's destiny. Inge says, 'Our training is as much about empowering locals to have a say and control over how tourism is developed on their island, as it is about providing first-class hospitality skills.' After completing the training at Masungi, many students go on to an internship at the island's renowned and sustainably minded five-star Nihi Sumba resort.

ABRAHAM PATH, MIDDLE EAST

Recognizing the power of tourism to tackle unemployment and gender differences, in 2014, The World Bank Group established the Abraham Path initiative. The idea is to develop a walking route spanning the Middle East – 1,000 km through Turkey, Jordan, Israel and the West Bank. Weaving between rural communities, the path connects travellers with homestays and small-scale enterprises often run by those with few income opportunities. So far, the project has generated 137 jobs, with 57 per cent of those going to women.

SEGERA, KENYA

In a very different setting, at Segera, one of Kenya's most sought-after luxury safari lodges, women are seeking equality in a male-dominated world. In 2019, twelve women were selected to take part in Segera's first All-Women Anti-Poaching Ranger Academy. The holistic, elite training prepares women to work on the frontline of conservation, not only providing the lodge with a vital skill but changing perceptions of the role of women in society.

It was designed in partnership with the International Anti-Poaching Foundation. Founder Damien Mander comments, 'Rather than recruit women to be rangers, we wanted to know what their dreams were. Who did they want to be? Being a ranger then became a vehicle to fulfil their destiny. Putting women at the centre of the programme's focus gave us the greatest traction in community development and relationships, and conservation simply became the by-product.'

GLOBAL HIMALAYAN EXPEDITIONS, NEPAL AND INDIA

A heart-warming example from the roof of the world is Global Himalayan Expeditions (GHE), which provides electric light to remote Himalayan villages. Paras Loomba, founder and former engineer, led his first 'electrifying' trip in 2014 when twenty tourists from all over the world hiked with solar micro-grids to a remote 1,000-year-old monastic community. Since then, GHE has electrified over a hundred villages that had never seen electricity before. Each expedition helps to install the micro-grids while engineers train villagers to maintain them.

Like so many off-grid communities, before GHE, households were reliant on often toxic (and carbon-intensive) kerosene lamps for light. Without electricity, rural life and traditions were dying quicker than ever as people migrated to towns and cities to be better connected. The simple act of providing renewable electricity means than villages have a better chance of survival. A knock-on effect from the treks is a burgeoning homestay industry if villagers want to earn extra cash from tourists.

ANGELINE LAMUNDE, SUMBA HOSPITALITY FOUNDATION GRADUATE, INDONESIA

Born and raised on Sumba Island, Angeline took part in the Sumba Hospitality Foundation's hospitality training and now works in the Front Office Department for Kempinski hotel in Bali.

–

How has Sumba Hospitality Foundation (SHF) changed your life?

Previously, I was a person who was afraid of dreaming, but now I am a person who dreams big! Before I never thought that I would work in a big hotel but because of SHF, I know that anything can happen.

Why do you think it's important for local people to be given this opportunity?

There are so many local people who are unable to pursue their education to high school. Sumba is a beautiful and unique tourism destination, and it is unfortunate if the community is only a spectator in its own region. Lastly, so that local people can compete and gain from tourism.

Do you think tourism has a positive or negative impact in your area?

Tourism has positive and negative impacts in my area. For example, the economic impact of tourism is usually seen as contributing to employment, better services, and social stability. It can increase available jobs and provide a higher quality of life for locals. Cultural education can also be improved. But tourism can also contribute to high costs of living in the community, increasing costs for local residents. One of the main threats to our island is that if tourism develops that is unsustainable, it will have a very negative impact on future generations.

What is your greatest hope for the future?

Firstly, I want to change the economic situation of my family and I want to help my siblings to continue their education to a higher level. Secondly, I hope that Sumba will become a sustainable tourism destination managed by young Sumba people, where the culture and natural beauty will be preserved for the next generation. Lastly, personally I hope to develop sustainable tourism on Sumba as taught by my schooling at SHF.

Giving Back

Rather than donating to a charity or handing over some secondhand goods, 'Impact Travel' is a new breed of trip that gives back. The term refers to adventures, expeditions or tours that put a positive social impact at the heart of the experience. Look for organizations or companies creating travel experiences that work a little harder, and address a tangible need, whether building community infrastructure or delivering medical supplies. Since each community and destination requirements vary hugely, these experiences should be tailored and bespoke; avoid any cookie-cutter approaches. Some platforms connect travellers with meaningful ways to give back, including Visit.org, Backstreet Academy and Airbnb Social Impact Experiences. US-based Impact Travel Alliance shares regular inspiration for how to travel to support poverty alleviation and equality.

CASE STUDY

RELIEF RIDERS INTERNATIONAL, MONGOLIA

Another adventure with purpose is Relief Riders International's horse-riding expeditions in Mongolia, India, Ecuador and Turkey (above). For over fifteen years, these intrepid adventures have delivered medical support to over 28,000 people by horseback. Some are so pioneering that they have been instrumental in influencing subsequent efforts by governments and the Red Cross. Founder Alexander Souri comments, 'When our riders experience themselves outside of their daily routine and travel intimately through one of the most vibrant countries on the planet, in a dynamic, heart-opening way, the effect is profound. It changes people.' And changing people, after all, is what travel is all about.

SOLIO CAMP, KENYA

If you like the idea of a little more luxury either side of do-gooding, The Safari Collection's Solio Camp has some bespoke impact travel opportunities. One eye-opening experience is donating and supporting the camp's annual eye clinic, which last year treated 1,110 patients and performed over a hundred life-changing surgeries. As many as 320,000 Kenyans are living with blindness, and 80 per cent of cases are highly curable. Solio hopes to bridge the gap, and guests can provide on-the-ground support in administrative roles in return for donations.

One patient, Karugu Migwi, describes finding out about the camp, 'I exhausted all my resources seeking treatment. It never worked; I had to go back home. It is a shame, and I cannot earn my living because of poor eyesight. When I heard the announcement over the radio about this medical camp, I told my wife: this is my last hope to recover my eyesight.' The procedure to remove his cataracts was a success and took only fifteen minutes.

FURTHER EXAMPLES

Throughout remote regions like Bhutan's Ura Valley (right) and India's Pindar Valley, hiking company Village Ways also facilitates community ownership. It works with each village to help them develop a tourism enterprise that the village then owns. Other successful examples include Kasbah du Toubkal in Morocco, which is now managed by the local Berber community, and Fordhall Farm Yurts in Shropshire, which is owned by 8,000 different local shareholders becoming England's first community-owned farm. In British Colombia, Spirit Bear Lodge is owned by the First Nation Kitasoo Xai'xais people with the mission of building capacity and employment opportunities for the people of Klemtu.

Prioritizing Community Ownership

When you start to break down where the money you spend on holiday goes, it quickly becomes apparent that tourism has a 'leakage' problem. Whether in chain hotels or restaurants or with an international tour operator, money spent in a destination often leaves the country to go into the hands of international businesses. It's a problem that's particularly prevalent in developing countries, which often need the revenue the most. Be part of the solution by injecting your money in locally owned business. In an ideal scenario, travellers would support lodges and tour operators that are either owned by local individuals or a whole community. But, that's not always feasible. In some destinations, there's not the local infrastructure to support all forms of tourism. In this case, it's essential to seek out responsible companies that are working towards models of community ownership, including funding community-based tourism projects, or helping a community create a viable tourist product.

STOP AND THINK:
BOOKING COMMUNITY TRAVEL

When community tourism isn't managed responsibly and sensitively, it can do more harm than good. Here are some things to consider before booking an experience:

- Community tourism mustn't be voyeuristic.
- If a community or indigenous people aren't benefitting from their stories and culture being shared, it's exploitation.
- Guides should always be local.
- If a situation wouldn't be comfortable at home, it isn't comfortable abroad either.
- Always ask permission to take photos and do so respectfully.
- If in doubt, ask: How would I feel in this situation if the tables were turned?
- Always follow local advice on interactions and cultural sensitivity.
- Never disrupt school lessons.
- Check operators have a child protection policy in place.

CASE STUDY

MYAING COMMUNITY BASED TOURISM AND INTREPID TRAVEL, MYANMAR

As part of a series of Community Based Tourism (CBT) projects, in 2019, Intrepid Travel launched the Myaing CBT Initiative in a rural pocket of Myanmar.

Homestays are illegal in Myanmar, making it hard for rural communities to contribute to or benefit from tourism. Looking for ways to overcome this challenge, the charity ActionAid introduced Intrepid to the villages of Myaing, a semi-dry zone in Myanmar's Magway Region where 1,000 residents rely on traditional farming. Income is scarce outside the agricultural season, and young adults are leaving the villages in search of work elsewhere, sometimes abroad. This has created a damaging level of insecurity.

After working alongside ActionAid to understand the needs of the community, Intrepid set out to develop a lodge and visitor experiences that would bring new economic prospects. Villagers contributed to the lodge designs, helped to build it, and gave permission for their common land to be used. It has been constructed using traditional materials, including bamboo, palm leaves and wood, in keeping with other properties and ideally situated so that visitors can cycle between each village. Visitors get to experience traditional cuisine made by locals and explore trails and footpaths. To date, the Myaing CBT has hosted 1,800 visitors.

CASE STUDY

RESERVA DO IBITIPOCA, BRAZIL

Over in Brazil, Reserva do Ibitipoca (above) is a 5,000-hectare private reserve in Minas Gerais: a vital wildlife corridor for an abutting national park. The reserve is home to a carefully restored luxury farmhouse hotel, but to ensure that the surrounding and largely neglected villages benefit from tourism, Ibitipoca has also established a new travessia. This walking or biking route meanders between several low-key guesthouses that have been restored with help from Ibitipoca. The centre of the whole project is Mogol – a village that has been repopulated thanks in part to tourist income.

BENSON KANYEMBO, WINNER OF 2019 TUSK WILDLIFE RANGER AWARD, ZAMBIA

Benson is the Law Enforcement Advisor for Conservation South Luangwa based in Mfuwe, Zambia. He's been committed to protecting wildlife and landscapes since joining Zambia's Department of National Parks and Wildlife in 1997. The Tusk Conservation Awards celebrate extraordinary people whose ground-breaking work with wildlife and communities in Africa might otherwise go unnoticed.

–

How has tourism had an impact on the work you do on the frontline of conservation?

It's had a positive impact by reducing poaching by providing employment to the local community in the area. Tour operators contribute money to pay the wages of the community scouts, and provide scholarships to school going children. Tourism also provides assistance to the human wildlife conflict mitigation programmes, the maintenance of access roads, and increases the GDP for the country. Lodges and bush camps act as eyes and ears of a conservation agency. Tourism is effective at creating awareness to appreciate the value of natural resources and the ecological processes that maintain them. The only real negative impact is that demand for bush meat is mostly driven by local people in employment, for example working in tourism.

Why is it important that the people you work with, your scouts, are local?

This is vital to enhance the relationship between the local community and conservation agency. When involved, communities have that sense of ownership and take full involvement in conservation. They also feel recognized for their importance in conservation and management of natural resources. The Community scouts are really crucial because they use local knowledge when doing their work – they are adapted to the climate, the landscape and know the local culture and traditions. It's essential that communities take responsibility for the management of their own natural resources.

Do you think tourism has a positive or negative impact in Zambia?

The positive impact is huge. It generates income, foreign exchange, employment and improves infrastructure and facilities. Environmentally, it enhances natural resources and helps to justify conservation. When it comes to culture, it can enhance cooperation and friendship. At a local level, it creates a market for local crafts and a platform to exchange knowledge – this strengthens the existing culture and institutions, raising their value.

That said, there are some problems; it increases the price of commodities, and it causes economic leakages – income often flows back to developed countries. There's also pollution, overcrowding and vehicles can disturb animals and habitats. Irresponsible tourism can also erode identities and tradition of place, and change moral behaviour.

What is your greatest hope for the future?

For Zambia's protected areas to be financially self-sustaining. For communities to embrace full participation and involvement of conservation programmes. And for Zambia to be one of the best tourist destinations in the world, making tourism the number one economic driver in the country.

Safeguarding culture

Over 40 per cent of travellers identify themselves as 'cultural tourists', which proves we're not all out to 'fly and flop' on holiday. This form of travel can be hugely positive for tourists, locals and heritage alike. Immersing yourself in a new culture forges connections, expands worldviews and increases understanding. It can also introduce us to new ideas and perspectives, which can make us better citizens – what can we learn from the past and how others live? This can be crucial to address social and environmental inequality.

Tourism can also fund the protection of heritage. It prompts communities to care for and take pride in their culture, and can even help people to create new expressions of cultural identity. Safeguarding culture can be vital for the health of our planet, too. Respecting and uplifting more traditional, rural ways of life can challenge urban migration and the destruction of landscapes.

But tourism is a double-edged sword. Last year, approximately 10 million tourists visited the most popular section of the Great Wall of China and nearly 20 million visitors explored Venice. As the Global Heritage Fund explains, 'When well-managed, tourism can protect these historic sites and provide an economic boost to surrounding communities. Conversely, poorly managed tourism results in congestion, saturation, commodification, and physical stress on infrastructure and on natural, cultural, and heritage attractions.'

Rather than following the crowds, we can use our travels to help lesser-known heritage sites or a marginalized culture. To do so, we need to seek historical sites that would be destroyed if not for tourism, and support social enterprises that help traditions to thrive. In this instance, social media can be a force for good. By delving deeper into local accounts and tips, it's possible to find off-the-beaten-track recommendations that aren't always in the mainstream. Local tourist boards and bloggers can be a good place to look, too.

The most responsible form of cultural tourism provides the funds and means for communities to protect their heritage on their terms. Culture is more likely to survive when it's living and breathing, and shouldn't be kept in a museum state for the sake of tourists.

CASE STUDY

AWAMAKI, PERU

An award-winning example is Awamaki in Peru. Located in the Sacred Valley of the Incas, Awamaki is a social enterprise that works to uplift the lives of women in remote Quechan communities. Founder Kennedy Leavens recalls her motivation, 'There were two challenges we wanted to help the women overcome. One was that their centuries-old textile culture was declining because its value was shrinking. The other was that women were being driven into cities for work, leaving villages to fall into ruin.'

Awamaki's two programmes, traditional crafts and rural community tourism have helped hundreds of women in remote villages surrounding Cusco. The Women's Cooperative Program helps weavers make more money from their textiles by collaborating with international designers and marketplaces. Meanwhile, the Sustainable Tourism Programme prepares communities to receive visitors for homestays and textile experiences.

Since 2009, trips including a half-day weaving workshop and four-day cloud forest natural dyes experience have hosted 5,000 tourists. Matching tourists with local families facilitates a two-way cultural exchange, ensuring that respect and mutual benefit are at the heart of every moment. To ensure the tours aren't intrusive, they are designed with the artisans and coordinated around pre-existing schedules, agricultural tending, and childcare.

Both programmes operate on an 'incubator model', with a focus on training and upskilling, so that women can establish independently and not be reliant on handouts. Kennedy is motivated to support women because of how they choose to spend money, 'on education and community improvement; so far, we haven't been proved wrong', she continues.

RIFT VALLEY, KENYA

Craft isn't the only form of tradition that tourism can help keep alive. In Kenya's Rift Valley, Sasaab Lodge is working with the local Samburu people to ensure their oral tradition thrives. The sounds and songs of the Samburu are as vivid as the wildlife that surrounds them – centuries of history has never been written down, but is recorded in song. To help protect this culture, Sasaab has partnered with Midi Minds, A Nairobi-based DJ collective, to record some of the Samburu's essential songs. Musicians and artists can buy samples, and the income is handed back the community.

NAYA TRAVELER, GUATAMALA

Bespoke cultural holiday specialist Naya Traveler also provides much-needed support for a centuries-old art-form in Guatemala. During week-long weaving tours, guests immerse themselves in craft workshops and learn how to backstrap weave among some of the few remaining artisans.

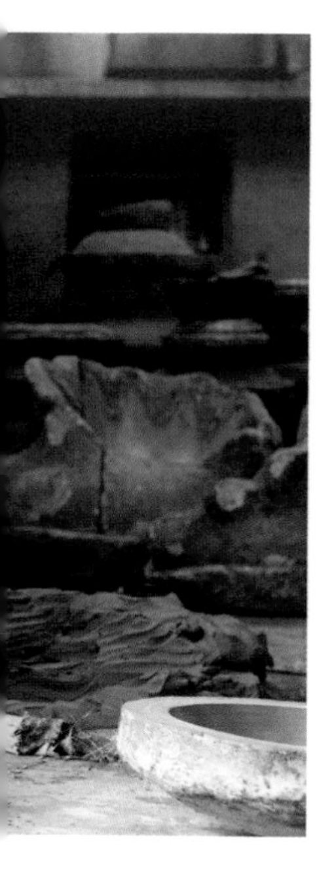

Helping Tradition to Thrive

All over the world, urban migration and the pursuit of economic progress threatens the survival of rural traditions and skills. As young people race to city centres in search of work and opportunity, rural areas experience mass depopulation, leaving an ageing population behind. With no one left to defend them, landscapes, ecosystems and communities often become vulnerable to exploitation and extractive industries such as mining, farming and construction. But tourism can help. When cultural experiences and projects are designed alongside communities, travellers can provide the supplementary income people need to continue a traditional or rural way of life. The money helps to ensure that the culture thrives. Local listings like Backstreet Academy in Nepal or Not on a Map in India link artisan and small-scale makers with tourists.

Protecting Heritage

The long-term survival of historical sites and heritage often lies in the hands of local people. But, communities need to be motivated to safeguard heritage and benefit from it, which is where responsible tourism can play a part. When tourism provides direct funds to a community surrounding a landmark, it not only offers financial incentive to protect it but also instils a sense of ownership and respect. This can have a ripple effect across a community, and even reverse urban migration as young people return home to benefit from their heritage. Responsible tour operators, travel designers and accommodations should be able to point travellers in the right direction. World Heritage Journeys, a partnership between UNESCO and National Geographic, is a platform promoting lesser-known cultural tourism projects that have a positive impact across Europe and Asia.

GLOBAL HERITAGE FUND, MOROCCO

A standout example is the work of the Global Heritage Fund (GHF). The US-based organization upskills local communities to restore and protect historical sites so that they can generate income. This work helps to tackle depopulation, which can happen not only due to dwindling economic opportunities but also climate change. Landscapes are becoming less dependable for farming.

One GHF project is working to protect abandoned granaries in the Anti-Atlas Mountains of southern Morocco. Often built into rocks and mountainsides, the granaries store irreplaceable documents and artefacts – many of which have been naturally preserved thanks to the arid conditions. With so many people leaving these rural areas, more of these sites are falling into disrepair and so the GHF is working with the Amazigh community to connect ageing master artisans with younger men and women. With new opportunities and creative industries within reach, the hope is that more young people will stay in the region.

Each year the GHF runs trips for anyone to explore and experience its work in Morocco, and all profits go back into the projects. The organization has joined forces with the world's first 'wandering hotel group' called 700'000 Heures – founded by transformational travel guru Thierry Teyssier. 700'000 Heures will create standout places to stay, and guests will enjoy exclusive access to the Global Heritage Fund's cultural sites.

DUNTON HOT SPRINGS, USA

The once-abandoned ghost town of Dunton, tucked into Colorado's snowy peaks, is proof that important heritage can crop up anywhere. Initially bought for real estate, owner Christoph Henkel fell in love with the estate's 200-year-old buildings, including the salon where Butch Cassidy carved his name into the bar. Local carpenters were brought in to rebuild each structure. The outcome was a luxury lodge, Dunton Hot Springs.

PORT GRAHAM, ALASKA

Further north is quite a different example of how tourism can benefit indigenous lives. When the First Nation Port Graham community was handed land rights to an ecologically rich part of the Kenai Fjords National Park, in the icy clutches of Alaska, they could have sold it off to extractive industries or large-scale tourism. Instead, they turned to sustainable tourism for its ecological protection. The community chose to work with Kenai Fjords Glacier Lodge – a responsible tourism operation that welcomes only 1,200 visitors per year. Via a small-scale lodge and low-impact wildlife trips, the lodge helps the Port Graham community make an income from the land while protecting it. It's a unique partnership that encapsulates why responsible travel efforts here are so crucial. The sanctuary is home to harbour seals, sea otters, river otters, black bears, bald eagles and oystercatchers. A complex web of life flourishes in a rare ecosystem where seawater and freshwater mix as glacial streams meet the sea. Endangered species, including humpback, sei and grey whales, rely on these rich waters.

HERITAGE WATCH, CAMBODIA

This non-profit organization based in Cambodia adds an integral part of the puzzle when it comes to protecting heritage. Rather than focusing on the tourists, it educates local school kids about why preserving temples and their archaeological artefacts is so vital. The project, Heritage for Kids, was launched in response to an increase in looting; an 8-m section of the nearby Banteay Chhmar temple was stolen one night. The charity is now working throughout Cambodia, providing resources for the government to roll out to schools and visitors alike. A series of sustainable development projects, including in Koh Ker and Banteay Chmmar, train local communities to benefit from and secure local archaeological sites and temples.

Supporting Indigenous Lives

Despite making up just 5 per cent of the world's population, indigenous people protect 80 per cent of the world's biodiversity. It's a staggering fact, especially when you consider that all over the world, indigenous knowledge and wisdom is suppressed in the name of modernity and progress. Supporting indigenous lives is a vital part of protecting our planet, and tourism can do its bit. It tends to take two forms; either the travel industry provides financial or in-kind support to ensure the protection of land rights, or it connects tourists with indigenous people that want to share their story. Sensitivity is vital, and no indigenous group should ever be subjected to tourism if they'd instead remain distant (see Stop and Think: Empowerment not Exploitation, page 88). Before travelling, always check advice and concerns with charities like Survival International. Canada has an entire indigenous tourism website promoting ethical experiences, and the recent banning of climbing Uluru proves that Australia is making headway bridging divides.

CASE STUDY

RUTAS ANCESTRALES ARAUCARIAS, CHILE

In Chile's wild Lake District 700 km south of Santiago, where rivers tear through the Patagonian rainforest, and allusive predators roam the slopes of volcanoes, lives South America's largest surviving indigenous group – the Mapuche. Historically alienated and persecuted, this ethnic group has fought against all the odds to maintain land rights and uphold a way of life that is symbiotic with respect for nature. While persecution is less than it once was, a battle against urbanization and modernity, and for respect continues.

Rutas Ancestrales Araucarias is a Mapuche owned and run tour operator based in Kurarewe, just 45 minutes from the glitzy tourist hub of Pucon. It was started when a few young Mapuche people didn't want to move to Santiago with everyone else, but couldn't find work in Araucaria. They turned to tourism. Initially, twelve families signed up to welcome tourists and show them their way of life – now thirty-two families take part alongside several shops, markets, a weaving workshop, cafés and homestays.

The aim is to give visitors (most of which are Chilean) a unique insight into Mapuche lives while providing a livelihood to the close-knit community. Experiences vary from an eight-day hike exploring the ancient tracks, traditional medicines and sustainable farming practices along the Pehuebche Route, to a one-day interpretative walk among the araucarias forest – home to the sacred *Araucaria araucana* tree.

Rutas Ancestrales Araucarias has been careful to develop a slow and mindful form of tourism. Guide and Communications Manager, Catalina Martí Puñoz says, 'The word tourist is very loaded; we prefer guest. We want to treat people like a guest, and we want them to act like a guest.' It's also essential that the community doesn't become dependent on tourism, so it's only a supplementary income, although it is a huge source of pride. Local chef Anita Epulef recalls how 'now the Mapuche cuisine is now valued throughout Chile, and even the world – hotel chefs come to ask me how to cook.'

STOP AND THINK: EMPOWERMENT NOT EXPLOITATION

With a rising interest in learning and transformative travel experiences, it's more important than ever for travellers to be mindful of exploitation. Consent is the key but that can be hard to judge when there's an off-kilter power-balance (is this travel's #MeToo moment?). Make sure a community or indigenous group always has control over what tourism they are offering, and if they want it at all – if you're unsure, quiz tour operators before booking. If possible, choose an experience designed by the community, rather than one that's been thrown upon them. Ask how much of your fee goes directly to tribe elders or community leaders – these are not 'human zoos' that other people should profit from. Think twice before joining a busload of others to visit a remote and rural community, and consider whether you would like to be papped before doing it to someone else. Always ask permission before taking photos. Check local regulations before entering indigenous reserves, read up on local concerns and issues, and never try to make contact with a remote tribe.

CASE STUDY

GUYANA

Guyana has won many awards for its cautious and sustainable approach to tourism. A crucial part of its success has been the work of the Guyana Tourism Authority's (GTA) strategic approach to indigenous tourism – building access roads and creating a 'tourism circuit'. Four indigenous communities have recently built ecolodges, which they own and operate. Brian T. Mullis, Director of the Guyana Tourism Authority, comments, 'We are making a concerted effort to scale up community-led tourism where many indigenous communities own and manage the tourism enterprises. This type of tourism provides visitors with meaningful immersive experiences and creates a sustainable income for the host communities.'

FURTHER EXAMPLES

Australia's Aboriginal owned and operated indigenous tours, like Bookabee, ensure that guests have an insight into Aboriginal Australian history and culture before taking part, and Wuddi Cultural Tours operates from a sensitively managed cultural centre. In both these examples tours are designed for Australian travellers as much as international visitors, which is vital in building bridges. Immersing travellers in two extremes, small tour operator Visit Natives works closely with indigenous peoples in the Arctic Circle and Tanzania – carefully selecting experiences that have the best possible impact.

Protecting nature

The most significant threat facing our planet is the destruction of natural habitats. If we're serious about fixing our environmental woes, this is where we need to focus our attention. Not only for biodiversity's sake but to store and absorb carbon from the atmosphere; according to NASA, forests store approximately 30 per cent of human's atmospheric carbon emissions

Forests, habitats, and landscapes are declining the world over as extractive industries like mining and industrial-scale farming encroach on wild places. In Brazil alone, a chunk of the Amazon rainforest the size of two football pitches is lost every minute.

But there is hope. Rather than decimating landscapes for short-term gain, the right kind of tourism can incentivize people to protect them in the long-term. In theory, it's the only industry with a vested economic interest in preserving the world's most awe-inspiring landscapes and captivating cultures. Tourism pounds will stop flowing if there's nothing left to see. In some cases, tourism is the last hope.

The most responsible travel experiences don't only save a species or protect land; they empower communities. When people living within and around a landscape benefit from tourism through jobs and enterprise, they are incentivized to safeguard it. Over time, visitor enthusiasm and praise generate local pride, and this helps to raise the value of the protected land and wildlife even further.

An organization that closely aligns with this thinking is The Long Run. With over forty members and growing, it helps tourism businesses to become first-class conservationists with culture and community top of mind. Managing Director Delphine King comments, 'The travel industry is at a crossroads: we can embed purpose and sustainability into our business operations, or we can watch our most important assets crumble under environmental and social exploitation. The world is full of beautiful rooms; people increasingly want beautiful rooms that protect the views they look over while supporting local communities.'

Contrary to common preconceptions, conservation-led travel isn't only about lavish ecolodges or far-flung corners of the world. It's equally important to respect and understand nature on your doorstep. Whether near or far, shoestring or luxury, here are a few ways to travel to protect nature.

Funding Protected Areas

All over the world, tourism plays a role in supporting protected areas. These areas may be privately owned reserves, national parks, or any defined area set aside for nature. Ensuring that ecosystems are protected, ideally in perpetuity, is critical given the climate and biodiversity crisis. The world's oldest rainforests in South America and Asia are some of the most significant ecosystems, given that they absorb so much carbon, but it doesn't stop there. Protected areas play a role in city centres (to clean air), and even under the sea; seagrass beds store more carbon per square kilometre than forests. Tourism can help by generating revenue for a national park or reserve upkeep, persuading local communities that nature is a valuable asset, and encouraging us all to foster greater respect for the natural world.

MISOOL ECO RESORT, INDONESIA

In the azure waters of Indonesia's remote Raja Ampat, Misool Private Marine Reserve was established by avid divers Andrew and Marit Miners after they witnessed the devastation caused to the area by dynamite blast fishing, shark finning, and mining exploration. The reserve now protects 300,000 acres of the world's most ecologically significant coral reefs – an area larger than all of New York City's five boroughs combined. WWF describes it as a 'species' factory thanks to the extensive 'no-take' fishing zone and the creation of a shark and manta sanctuary.

Thanks to Misool's work alongside the government and local community, the reserve is a pristine wonderland for tourists that stay and dive from the small eco-resort. Oceanic manta rays visit underwater cleaning stations, sandy beaches are home to nesting hawksbill and green sea turtles, and blacktip reef sharks breed in shallow lagoons. In 2017, the reserve achieved Mission Blue Hope Spot status, proving its critical importance to ocean health.

BASECAMP OULANKA, FINLAND

Just below the Arctic Circle, Basecamp Oulanka's sustainably minded lodgings offer visitors a chance to spend the night in Finland's wild Oulanka National Park (opposite, above). Northern Lights, bears, wolverines, and Artic snowscapes create unforgettable experiences, but it's Oulanka's revival of fragile ecosystems that is remarkable. Funds from adventure tourists are used to protect the world's largest forest – the Taiga. Through its WildOulanka Foundation, fees from Basecamp Oulanka help to rent land from a local community (the Kuusamo Forest Common) to create a 1,000-hectare wildlife corridor between Oulanka National Park and Paanajärvi National Park in Russia. The foundation has helped to establish a local 'bio-economy' so that the community is reliant on sustainable forestry and low-impact tourism rather than logging. Without this additional protection, species like wolves and elk may not have the landscape they need to survive.

CHUMBE ISLAND, ZANZIBAR

One of the first places in the world to embrace this model is the tiny island – only 11 km by 300 m – of Chumbe Island Coral Park, 8 miles off the coast of Zanzibar (opposite, below). The coral reef sanctuary is home to 90 per cent of East Africa's hard coral species and over 400 reef fish species. Rare wildlife protected on and around the island includes the coconut crab, green turtle, Ader's duiker (antelope) and humpback whale. Low-impact bandas (thatch villas) are dotted along the island's white sands, and have been designed with sustainability top-of-mind, making this one of East Africa's most pristine nature experiences.

Guests that visit the island's pioneering ecolodge aren't the only ones to benefit; since 1995, over 9,000 local people including school kids, teachers and government officials, have travelled to Chumbe for conservation education. Activities include swimming lessons (many of the locals can't swim), snorkelling trips, and taking part in conservation activities at the island's dedicated education centre. Local fishers have been trained as park rangers so they can still generate an income.

WILDERNESS SAFARIS DAMARALAND CAMP, NAMIBIA

Among Namibia's red ranges and shifting sands, a pioneering partnership between the Torra Conservancy and Wilderness Safaris proves that tourism and community can work together for the benefit of people and wildlife alike. Since the 1970s, Damaraland has been home to indigenous groups ousted to the far-flung desert outpost during the apartheid. Harsh desert conditions, and zero land or water rights, made prospects bleak and poaching high. Recognizing this plight, in the mid-1990s, Wilderness Safaris created a joint venture with the community to establish Damaraland Camp where visitors can experience life in the desert, which is home to elephant and black rhino. From virtually zero employment, now most young community members work in tourism, trained by Wilderness. Consequently, the wildlife and ecosystem are thriving; more than 350,000 hectares is protected in the long run.

Restoring Ecosystems

Rather than merely protecting wild landscapes, tourism can also help to restore them. As we become more mindful of our impact on nature, rewilding is an increasingly popular phenomena world over. It refers to the act of putting the land back to its natural state – reintroducing native species and removing livestock so that wild flora and fauna can thrive. Tourism is often used as a vehicle to restore ecosystems because travellers are more likely to pay to see a healthy habitat than a damaged one. Creating new landscapes from abandoned quarries or intensive farms is also a great way to lure tourists away from overcrowded national parks or hotspots. Rewilding Experiences lists relevant places to stay and tours across South America, and Rewilding Europe works with local partners to promote the continents most critical ecosystem restoration projects.

CASE STUDY

ROUTE OF PARKS, CHILE

Established by pioneering conservationist Kristine Thompkins (with her late husband, Doug, she set up the Thompkins Foundation), Chile's new 1,700-mile Route of Parks connects over twenty-five years' worth of strategic land-acquisition for the sake of conservation. Devastated by what they saw of mass extinction, often due to neglect or over-grazing, Kristine and Doug gradually bought over 28.4-million acres of ranches and private land in Chile to create wildlife corridors and return the wilderness to its natural state.

Five new national parks now connect the more touristed hotspots of Punta Arenas and Torres del Paine with remote areas like the Aysén region and Pumalín. Thompkins' hope is that it will appeal to everyone, whether staying in a luxury lodge or wild camping. The aim is that local communities embrace the opportunity to develop small-scale, low-impact tourism businesses. Over sixty local communities could benefit from the parks, and the Amigos de los Parques (Friends of the Parks) ensures that Chileans utilize and feel pride in the revamped national park system.

CASE STUDY

MASUNGI GEORESERVE, THE PHILIPPINES

Forty-five kilometres east of Manila is a unique limestone region home to prehistoric karst formations including caves that plummet to the water-level and natural sculptures that tower above the indigenous rainforest (opposite). By the 1990s, much of the landscape has been devastated by illegal logging and quarrying. Recognizing its ecological significance, the government stepped in to create an action plan for its protection, and an unlikely conservation hero rose to the treetops.

Previously desolate, an area of wild rainforest and gnarly rocks that host native Filipino creatures and critters, is now home to day-trippers clambering over giant cobwebs, hanging ladders, wire pods and rope courses. Since 2015, the Masungi Georeserve has used adventure and nature-seeking visitors to fund the restoration of the precious ecosystem. The unusual and spectacular structures, created with biomimicry in-mind (mimicking nature's structures), create a low-impact adventure for hundreds of people a day.

The adventure trails were designed with nature and conservation in mind. So far, over 40,000 native trees have been planted, and 400 species identified – local school kids now draw civet cats rather than tigers. Former poachers are now park ranchers, and more than 100 local people have gained employment. In 2017, the government expanded the area under perpetual conservation from 400 hectares to almost 3,000 hectares.

CASE STUDY

TAHI, NEW ZEALAND

On the shores of New Zealand's mild 'Far North' is Tahi, a sanctuary where nature has been reawakened. Eight-hundred acres of estuaries, wetlands and forest meet golden sands and wild Pacific surf. On the dunes, it's possible to find the remains of eggs from the sadly extinct moa, while the waves beyond provide year-round surf. Whether experienced on horseback, kayak, foot or bike, nature consumes everything.

When Suzan Craig bought the land in 2004 it was a run down, derelict cattle farm. The reserve seen today is the result of fourteen years of hard conservation graft. The landscape was once a fertile haven where Maori culture thrived. But when European settlers arrived in the mid-1800s, they transformed the landscape, cutting the native forests for building material and fuel, introducing farming and a host of pest species that decimated the native wildlife. Wetlands were drained, the birds disappeared, and the land lost its soul.

'Before the arrival of people' explains Craig, 'New Zealand was a land of birds. But with human occupation came a host of mammalian pests, including stoats, weasels, rats, cats, possums, pigs, goats and rabbits. Pest control is now vital. Here at Tahi, sustainability is a living philosophy. We consider ourselves to be kaitiaki (custodians) of this land and we're passionate about our restoration efforts, that allow native species to return.'

FURTHER EXAMPLES

In Cornwall, UK, the quirky campsite Kudha is encouraging nature back to an abandoned quarry. Owner Louise Middleton is using profits from Danish cabins and tree-tents to bring back nature including a willow wood and low-lying plants for butterflies. Further north, Cairngorms Connect is one of the UK's most ambitious rewilding projects. Supported by the luxury travel group Wildland, with funding from the Endangered Landscape Programme, the project is working to restore over 600 km^2 of land.

STOP AND THINK: ARE SOME PLACES BEST LEFT UNVISITED?

Where travel doesn't aid conservation or support communities, then are those places better left alone? One example that causes much debate is Antarctica. There's something about the earth's icy extremity that is irresistible to humans; a mysterious half-land that has gobbled up hardy adventurers and left others in utter awe. Its ultramarine icebergs are so sought after that the region has recieved up to 50,000 visitors a year. There are a few responsible travel arguments: marine biologists and scientists occasionally hitch a ride on cruise ships; it can help people to grasp the reality of climate change (although many ships avoid explicit mention of the topic for fear of 'conflict'); it rallies consensus that the region needs protection. But is that enough? The average carbon footprint of each passenger visiting Antarctica is 5 tonnes of CO_2, and soot or black carbon from ship exhausts can build up on ice and increase warming. The International Association of Antarctic Tour Operators (IAATO) regulates how tour-operators behave, but signing up is voluntary, and the ship numbers keep on rising. How much is too much?

Saving a Species

In 2018, the World Travel and Tourism Council (WTTC) discovered that wildlife tourism contributes 5.2 times more revenue than the illegal wildlife trade (poaching, trapping and snaring to sell wildlife for medicines, furs or food). When governments, communities and businesses realize that wildlife is worth more alive than dead, they are more likely to take a conservationist approach. Tourism is providing the incentive to support endangered species world over from Macurian bears in Italy to manatees in the United States. For some species, this is the last hope. According to John E. Scanlon, Special Envoy for Africa Parks, the WTTC Declaration to fight the illegal wildlife trade (known as the Buenos Aires Declaration) is a good place for travellers to find out more. He says, 'Over a hundred companies have signed, and others can join. Travellers can support good tourism operators who are supporting local conservation initiatives, make sure they never buy illegally sourced wildlife, and report suspicious activities they may see to staff.'

STOP AND THINK: ANIMAL WELFARE

Animals all over the world are abused for the sake of 'tourism' – whether drugged for photo opportunities, abused to encourage performance, or mistreated in zoos and so-called 'sanctuaries'. The Association for British Travel Agents (ABTA), updated its Wildlife Tourism Guidelines in 2019 to make this clear. Unacceptable behaviours include any tourist contact or feeding of great apes, bears, sloths, crocodiles, or alligators, ostrich riding, and feeding or walking with wild cats. Also deemed unethical is any photo or holding opportunity with a wild animal where it can't move away at any time and feeding or contact with animals in the wild. Elephant rides, shows, or bathing are also on the list. Whale and Dolphin Conservation recommend that visitors join responsibly minded whale and dolphin watching trips rather than interacting with them in captivity (any responsibly minded travel company will have pulled out of Sea World by now). In 2019, an undercover report by consumer magazine *Which?* found that nine out of ten industry giants, including Trailfinders, Virgin Holidays and Expedia, were still selling inappropriate animal experiences.

CARPATHIA, ROMANIA

Another giant in need of protection is the European Bison. Most at home roaming Romania's Carpathia mountains and valleys, the species was hunted out of the wild in the late eighteenth century; since then they've been on the International Union for Conservation of Nature's (IUCN) 'vulnerable species' list. Over the last five years, European bison have been gradually reintroduced to Romania as part of broader rewilding efforts. Such reintroductions are vital for the creation of what not-for-profit Foundation Conservation Carpathia (FCC) hopes will become the 'Yellowstone of Europe' – over 200,000 hectares of protected forest. The success of the bison and the forest depends largely on sustainable – low-impact, high-value – tourism since the surrounding population needs the incentive to make it work. Travel companies like the European Nature Trust and the European Safari Company are working hard to make 'bison' safaris work for visitors, bison and the local population alike.

BORANA LODGE, KENYA

In many parts of Africa, private tourism businesses have become fundamental in the survival of a species. Perched above Laikipia's infamous wilderness – a savannah that is home to buffalo, rhino, lion, and elephant, dissipating into the distant jagged contours of Mount Kenya – Borana Lodge and its 32,000-acre conservancy is one of Kenya's most sought-after destinations (opposite). In 2014, fences between the neighbouring Lewa Conservancy and Borana were removed, creating Kenya's largest continuous rhino habitat; 93,000 acres is now home to over 200 rhinos and hosts 14 per cent of Kenya's black rhinos. Michael Dyer, Managing Director of Borana, comments, 'Providing habitat for rhino is a huge undertaking. The cost of securing land (male rhinos need 3000-acres of home range) and providing 24-hour security 365 days of the year against poaching is high. We couldn't do it without tourism.'

AKAGERA NATIONAL PARK, RWANDA

John E. Scanlon recommends this thriving ecosystem as proof of the power of tourism. He says, 'Just over a decade ago the Akagera National Park in Rwanda was largely poached out, with little or no tourism, generating no local benefit. It was almost lost. Fast forward ten years, through a partnership between African Parks and the Rwanda Development Board, the "big five" African animals have been returned to the park, poaching has ended, and three tourism facilities, from five to two star, have been developed. In 2019, over 50,000 tourists visited the park, half of them Rwandan, generating many decent local jobs. The park is today thriving ecologically, financially and socially, and is generating 90 per cent of its operating costs.'

CEMPEDAK, INDONESIA

If those living within or surrounding a landscape have ownership over its future, conservation is more likely to stick. Some of the most inspiring examples include tourism's role in converting hunters into wildlife guides, which can have a hugely positive impact on the future of a species. In Indonesia, eco-resort Cempedak has worked closely with the surrounding indigenous sea-faring nomads Orang Laut for many years. Thanks to frequent talks about conservation and the fate of the hunted dugong (a vulnerable sea mammal) one local leader and fifth-generation dugong hunter, Musa, has changed his ways. He's now paid by Cempedak to guide tourists to the dugong rather than kill them.

SEE TURTLES, COSTA RICA

A similar approach has been taken in Costa Rica, where conservation not-for-profit SEE Turtles uses funds from visitor tours to educate people living around turtle nesting 'hotspots'. Over 100 local teachers and community leaders have received turtle conservation training, and 2,000 students have been on field trips to the turtle nesting beach. Via turtle conservation, local businesses have made over $300,000 and over 1.7-million hatchlings have been saved.

RHINOS WITHOUT BORDERS AND LIONSCAPE

Another way the travel industry can help save a species is through collaboration. Competitors and Beyond and Great Plains Conservation raised over $5-million for their joint Rhinos without Borders project. With the help of the Leonardo DiCaprio Foundation, several safari outfits have formed the Lionscape Coalition to work on a cross-continent lion conservation project.

STOP AND THINK: WILDLIFE AT WHAT COST?

In Central Africa, the plight of mountain gorillas was so dire that experts expected the species to be extinct by 2000. Today, there are more than 1,000 roaming the forests of Congo, Rwanda and Uganda thanks to national parks that were created in the 1990s. But, it's not necessarily a happy ending for everyone. To create a national park for the gorillas, one of Africa's oldest tribes, the Batwa, was pushed out, sometimes violently, from lands they had subsided on for centuries. The plight of the tribe has been a reminder of this harsh reality ever since – without land rights, many groups have fallen victim to disease, alcoholism and abusive industries. Some travel companies have done more than others to ensure their survival – Volcanoes Safaris, for example, has worked hard to create community centres and income opportunities for the Batwa. Some now live peacefully alongside the park, and also benefit from tourism. But, the more dire situation for others is a reminder that conservation can come at a human cost. While sometimes, there's no perfect solution, protecting wildlife should work in harmony with people that depend on the land, rather than against them.

Contributing to Citizen Science

Citizen Science invites members of the general public to help scientists and conservationists collect data from the natural world. It's one of the most effective ways that tourism can directly contribute to vital conservation efforts. This might involve conducting wildlife surveys, monitoring a species, or sending photos to an online collaborative effort. Thanks to a rise in travellers looking for more purpose-led experiences, opportunities are more varied than ever. When travelling, you can either choose an expedition with citizen science at its core or take part in a citizen science project on the side. In total, there are more than 3,000 citizen science projects in action, including Bioblitz and iNaturalist that invite anyone to upload wildlife observations.

CASE STUDY

BIOSPHERE EXPEDITIONS

Since 1999, UN Environmental Programme member Biosphere Expeditions has been on a mission to create a more sustainable planet via meaningful travel experiences. Its first expedition took a team of ordinary people to the Polish Carpathian mountains to study wolves, which ended up being instrumental in a wolf hunting ban. Other successes include expedition data being used to help form the Saylyugemsky National Park in the Altai Republic of Russia and the creation of a new national park in Ukraine.

Today, Biosphere Expeditions world over connects travellers to locally led wildlife research and conservation efforts. In the United Arab Emirates, for example, the expedition heads into the iconic sands of the Arabian Peninsula to work alongside scientists monitoring the Arabian oryx, Gordon's wildcat and other flagship species within the Dubai Desert Conservation Reserve. In Kyrgyzstan, travellers get a unique opportunity to understand the inner workings of conserving the allusive snow leopard in the remote Tien Shan mountains. Each day the expedition works with a small international team from a mobile camp base to look for tracks, kills, scats and setting up camera traps.

Drawing on twenty years of experience, Dr Matthias Hammer, founder of Biosphere Expeditions, emphasises that people must choose volunteer projects wisely. Aside from carefully checking for qualified staff, where your money goes, and animal welfare guidelines, Hammer says integrating local communities and their needs is vital. He says, 'If you are thinking about volunteering, make sure you understand a project's relationship to the local community and that the organization is properly embedded with local efforts and people. For example, does the community benefit, have they given consent for work to be carried out, how have they been involved? Is there training for locals, scholarships, capacity-building, education, etc.?'

FURTHER EXAMPLES

There are other opportunities all over the world. Wildsea is a collection of seafaring and citizen science opportunities around Europe's abundant shores, and the Oceanic Society offers snorkelling expeditions. Meanwhile, Earthwatch takes adventure-seekers into lesser-visited parts of the Amazon to develop strategies to combat poachers, and Much Better Adventures Belize trips get travellers helping Reef Conservation International remove the invasive lionfish species from Caribbean seas. In Mexico, Sustainable Travel International has launched NEMO (Natural Environment Marine Observers) to involve travellers in protecting Mexico's vast Mesoamerican Reef.

Contributing to citizen science doesn't have to involve an epic expedition. Throughout the UK, organizations like the Wildlife Trusts and Royal Society for the Protection of Birds (RSPB) list where you can see lesser-known wildlife and how you can contribute to their survival. For example, the RSPB asks puffin-watchers to submit photos they take to help monitor numbers and locations. Based in the United States, The Nature Conservancy uses data from hotels and tourists to prove that healthy coral reefs generate $36 billion each year, which incentivizes governments to protect them.

REGENERATIVE TRAVEL

Rebuilding places

Converse to overtourism, 'undertourism' describes places that desperately need tourists. These are destinations suffering from a severed reputation, or places torn apart by a natural or humanmade disaster. When it comes to travelling responsibly, visiting a destination hampered by 'undertourism' is often the cream of the crop; these are the places where tourist pounds go furthest and can help to rebuild lives.

Examples of areas impacted by natural disasters include Nepal after the devastating earthquake of 2015 when tourism collapsed overnight, or Haiti, Mozambique, and Thailand that have also struggled with life-changing events. Tragically, the list is only likely to grow with the onset of the climate crisis.

Civil unrest, conflict, terrorism and disease can also wipe destinations off our wanderlust radar. Annual visitors to Egypt have plummeted almost 10 million since the 2016 uprising, and Sri Lanka's tourist numbers shrank by 57 per cent after the bombings in 2019. Coronavirus has taken a huge toll on tourist bookings in the worst-affected regions. It can take decades for a country's reputation to be repaired and tourists to return; in the meantime, hundreds of people's livelihoods are put on hold. While the media is quick to report a crisis, it's usually slow to deliver news of recovery. Intrepid adventurers are often the first to return and they can often play an role in restoring faith and instilling hope.

Expert Peter Singleton, Lecturer and Researcher at NHL Stenden University of Applied Sciences in the Netherlands, comments, 'The major element in rebuilding a destination that has declined in tourism flow is to choose to visit. This requires the objective, open-minded stance which allows for conditions on the ground being safer than the image of risk would suggest. Often the image does not correspond to reality, so this requires a different mindset on the part of the tour operator and tourist.'

Of course, acting responsibly isn't jumping on a plane the moment you hear about a crisis. Emergency services and experts need time, space and generous donations to do vital recovery work first and foremost. The role of the responsible traveller comes in later on – when an emergency response comes to an end, and people need to get on with their lives and welcome tourists back. Rather than merely contributing to an economy, it's in these destinations that tourism can help to rebuild lives, restore nature, heal history, instil hope, and bring places and people back from the brink.

Recovering from a natural disaster

There are three ways in which travellers can help places recover from a natural disaster: volunteering on the ground where immediate help is needed; supporting areas unaffected by the crisis; healing reputational loss once recovery is complete. Since no two disasters or destinations are equal, it's vital to research before booking any trip. It's best to find out what people are saying on the ground via local news sources and speaking with the tourist board. Whether booking with them or not, turn to reputable responsible travel operators for additional information. Responsible Travel, Much Better Adventures and Intrepid are a few that produce measured content in the wake of a crisis.

CASE STUDY

NEPAL

It's a similar tale in Nepal. Before the 8.1 Richter-scale Gorkha earthquake devasted much of the Kathmandu region, tourism provided a staggering 500,000 jobs (partly due to the labour-intensive nature of trekking). For a couple of years after the quake, tourist arrivals were down 32 per cent. While volunteer tourism of any kind requires scrutiny (see Stop and Think: Voluntourism, page 72), in the case of Nepal, experts believe it was vital in the destination's recovery. The Pacific Asia Tourism Association worked alongside tour operators, communities and the governments to create a Nepal Tourism Rapid Recovery Taskforce that recommended using volunteer tourism to fund recovery and welcome visitors.

Equally crucial in Nepal's recovery was to keep visitors coming to largely unaffected regions. This included the popular Annapurna Circuit trekking route above Pokahara (opposite, below) – guesthouses, porters and guides all suffered as a consequence of the earthquake, even though structurally, it had no impact on the experience they were selling. In response, Responsible Travel worked with tour operators to curate trips that combined trekking with volunteering efforts, with 30 per cent of tour fees going towards earthquake aid.

CASE STUDY

DOMINICA

In September 2018, Hurricane Maria swept through Dominica ravaging the homes, communities and nature in its path. Barely a tree was left unscathed as the lush forest lost its emerald tones, and most tourism businesses were flattened in a matter of hours. Yet, the population did what so often humans do best; rebuild and cooperate for survival.

Jem Winston, Founder of 3 Rivers & Rosalie Forest Eco Lodge in Dominica (opposite, above), recalls how vital tourism was to help rebuild the island, 'Our recovery has been solely as a result of tourism . . . After the hurricane, guests volunteered their time in aiding not only our ecolodge, but the surrounding villages as well. Moreover, one of our partners, Reponsibletravel.com, suggested and helped us implement a voluntourism package. This resulted in many visitors, with some very useful skills, paying for their stay, which we really needed.' The island is now working towards becoming 'the world's first climate-resilient nation', helped by support from The Clinton Foundation and Price's Trust.

PUERTO RICO

After Hurricane Maria hit Puerto Rico, many questioned whether it was ethical to luxuriate in an all-inclusive resort while much of the population remained homeless. Yet, speaking to National Geographic, Brad Dean, CEO of Discover Puerto Rico, argued that 'Locals understand the immense contribution that tourism provides the economy of Puerto Rico. Visiting is not insensitive; it's benefitting the economy and providing Puerto Ricans with jobs in the tourism industry and beyond. We've always been known for our warm hospitality, and now is no different.'

STOP AND THINK: CLIMATE CRISIS TRAVEL

Every time there's a natural disaster somewhere loved by tourists, social media feeds fill up with devastating statistics, appeals for help and calls to travel there. While this response often has best interests at heart and contains some truth, maybe we're better off taking stock and considering the bigger picture. What are these natural disasters telling us about the state of our planet? Should we respond like its business as usual or should we make more fundamental changes to our behaviour? If we do take stock, we might choose to shy away from the great irony of releasing tonnes of carbon into the atmosphere to help those affected by climate-related natural disasters. Instead, we could dig a little deeper into our pockets and donate to conservation charities on the ground looking for long-term solutions. Or, we might feel a surge of motivation to reduce our carbon footprint. We may also change how we travel: since a five-year drought worsened the Australian fires, then, if we are going to travel to Australia, how can our travels help to prevent drought? We need to support businesses and people protecting and restoring ecosystems that are more naturally resilient than intensive farming. Going one step further, how can we learn from natural disasters to improve destinations in the long-term? Rather than merely seeking recovery, we could aim for relief efforts to be regenerative and transformative.

FURTHER EXAMPLES

Dominica, Puerto Rico and Nepal aren't isolated cases of responsible tourism joining the road to recovery. In Japan, after 2011's tsunami, locals formed a Blue Tourism campaign group to promote the region. In 2004, parts of South East Asia unaffected by the tsunami, like The Philippines, Goa, and Malaysia's east coast, pleaded with tour operators to keep sending visitors fearful that an 'economic tsunami' may follow. More recently, the not-for-profit website run by California, Oregon and Washington, the West Coast Tourism Recovery Coalition, helps to provide travellers with real-time information regarding wildfires, to dispel rumours and keep visitors safe.

In Australia, there's a similarly cautious approach; the real upshot of 2020's catastrophic events is yet to be seen. Meanwhile, responsible travel operators are doing everything they can for those affected. One example is specialist tour operator, AAT Kings, which has launched a series of Bushfire Relief Giveback Experiences in partnership with Empty Esky. The grassroots campaign inspires people to visit the Australian towns affected by the bushfires to inject money back into the community. Taking visitors through neighbourhoods affected by the fires, 100 per cent of profits go to support small businesses and farmers most in need.

Climbing out of conflict

It sounds improbable, but sustainable tourism can help to facilitate peace. In 2016, a research report conducted by the Institute of Economics and Peace and the World Travel and Tourism found that 'Countries with a more sustainable and open tourism sector are more likely to enjoy higher levels of positive peace in the future.' The right kind of tourism – that's constructive and ethical, rather than voyeuristic 'dark tourism', where people rush to war zones – can break down cultural and cross-border barriers, fund memorials for peace, and incentivize cooperation between groups traditionally in conflict. It can also instil hope and bring economic prosperity for places reeling from humanmade disaster. Adventure tourism is often the quickest form to return; activities like hiking trails or cycling routes require less infrastructure and put tourist income in the right places – for example, locally owned guesthouses. Or, travel with a respected operator like Wild Frontiers that puts local needs first while challenging traveller preconceptions.

CASE STUDY

PEAKS OF THE BALKANS

Notions of peace can also lie in something as simple as a hiking trail. In 2013, the transnational 192-km Peaks of the Balkans route was recognized by the Tourism for Tomorrow Awards for opening both minds and borders. The trail uses mostly old shepherds paths, winding through deserted high alpine mountains up to 2,300 m above the sea level. The route spans three formerly war-torn countries – Kosovo, Montenegro and Albania – providing an income for rural communities and proving that cooperation is the future. One of the hardest aspects of creating the route was persuading border police to allow free access.

CASE STUDY

ALBANIA

Tourism is providing hope for each of these countries. Albania, once considered a poverty-stricken mafia stronghold, where post-Communist rioting took hold throughout the 1990s, is an increasingly sought-after destination. Adventure travellers were the first to trickle in, drawn to its alluring coastlines, mountain trails, World Heritage Ottoman architecture, and some of Europe's most exquisite mosques. As Europe's fourth most impoverished country, inequality is still omnipresent, but responsible tourism can spread income into rural areas by rewarding self-starting entrepreneurs. Sustainable tourism is particularly suited to offering opportunities to marginal communities and women, without forcing them to forgo traditional ways of life.

Tourism has another subtler regenerative role to play, too. Sam Bruce, Co-founder of Much Better Adventures, explains, 'We're always amazed at how intrepid Eastern Europe feels. In places like Bosnia, no one talks about the recent, brutal history. Yet, when asked sensitively, our local guides are pleased to open up. They want to share their story. It's a missed opportunity for those that don't. Surely building greater understanding between different people and cultures is an important part of adventure travel.'

CASE STUDY

RWANDA

In central Africa, a study conducted by Swiss Peace found that gorilla tourism was vital in facilitating cross-border collaboration. The common interest of gorilla conservation in the Virunga-Bwindi region of Uganda, Rwanda and the Democratic Republic of Congo helped the former enemies work together to create a transboundary park.

In Rwanda, joint ventures between tourism companies and local communities helped the assimilation of vulnerable groups in a post-conflict society. For example, The Sabyinyo Silverback Lodge in Volcanoes National Park was built by a wildlife safari company but is owned by the Sabyinyo Community Lodge Association (SACOLA) – a community trust representing 6,000 low-income Rwandans from a nearby village.

FURTHER EXAMPLES

If you are travelling to a destination with a troubled past, seek experiences that focus on healing divides and enhancing visitor understanding. Operating throughout Colombia, Brazil, Sri Lanka, Guatemala, Mexico and Belize, tour operator Justice Travel helps visitors get closer to issues on the ground. Through varied experiences, from street art tours in Mexico (above) to treks with ex-FARC guerrillas in Colombia, it aims to 'move beyond racist nationalism, beyond media-induced apathy, and beyond our own perceived boundaries to bring about a brighter and more just future'. In Ireland, the Belfast Free Walking Tour leads visitors through 'The Troubles' with ex-political prisoners from both communities. In Palestine, the Siraj Centre Palestine was born out of frustration that the country was only known for holy sites and violent conflicts, and gets tourists out into hiking trails and homestays.

STOP AND THINK: TAKING RISK

Have we forgotten how to take personal responsibility for risk? The Canadian government's travel advisories page recommends that travellers 'exercise a high degree of caution' in destinations including Indonesia, India, Nepal, Peru, Philippines and the United Kingdom. Meanwhile, the United States has a blanket warning: 'As terrorist attacks, political violence (including demonstrations), criminal activities and other security incidents often take place without any warning, US citizens are strongly encouraged to maintain a high level of vigilance when travelling abroad.' When on holiday, often the last thing we want to do is contemplate risk; we barely want to make any decisions, let alone be confronted with social tensions. But when we crave adventure and want to understand the world a little better, taking an element of risk is part of the package. Only when we step out of our comfort zones can we challenge preconceptions and distil greater understanding. Arguably, the more we step out of them, the most enlightening experience we're likely to have – senses sharpened and eyes wide open. Of course, everyone has their boundaries when it comes to taking risk. But, it's worth considering that in an age where overtourism is sweeping across much of the world, sometimes championing underdogs – for example, Turkey, Iran, Pakistan, Sri Lanka, Sierra Leone, Nicaragua – are our best chance at an authentic and meaningful experience. It can do wonders for these destinations, too.

Improving cities

By 2050, two-thirds of the world's population will live in cities. At the moment, our urban epicentres make up just over 3 per cent of the world's landmass but account for three-quarters of global pollution and 70 per cent of energy consumption. As such, making cities more sustainable is a crucial part of our fight against the climate crisis; not only for the planet's survival but also for global health and happiness.

Tourism is intricately woven into the fabric of many cities. It can trigger rapid urbanization but also can help to halt it by providing livelihoods in rural areas. Either way, the needs of residents and visitors are often in sync. What makes a city a good place to live – clean air, affordable transport links, lively culture, treasured heritage, accessible open spaces, and services to suit a range of budgets and needs – also makes a city a great place to visit. While the needs of residents must always come first, tourism can facilitate progress.

In theory, cities are ripe for sustainable living. By living in such proximity, city-dwellers should be more efficient, both economically and environmentally. The reality, however, is often quite the opposite. Many parts of the world have experienced urban growth at an unprecedented rate, and cities haven't had the time to adjust thoughtfully or sustainably.

Instead, urban areas are often associated with inequality and pollution. By 2030, the UN estimates that over 2 billion people in the world will be living in slums. Even in the developed world, economic endeavour favours speed over sociability. So most major cities have been sliced and diced by petrol-guzzling arteries to the detriment of the environment.

Recognizing this balance of tensions and hope can help us to understand how we as travellers can improve cities. By visiting the world's more 'liveable' cities – those championing green technology, innovative solutions and social equality – we can incentivize and support progress. Where this is not possible, we can support grassroots environmental and social projects funding the changemakers seeking improvement.

We can also travel to improve our home cities by bringing back with us news ideas and innovations to create a ripple effect of change in our own neighbourhoods.

Greening urban spaces

Adopting a more eco and nature-friendly approach to city-living requires significant investment. Tourism can play its part, particularly when destinations develop sustainability policies that put nature first. Some recent examples include Singapore, which has increased its green cover by 10 per cent in the last twenty years, and Vienna, which has created its first 'climate-adapted street' with gardens, spaces for bikes, and trees. Sweden's second city, Gothenburg, has topped the Global Destination Sustainability Index two years running. As mentioned in 'Where to stay' (pages 54–65), by choosing green hotels and places to stay helping to promote biodiversity and innovate with green tech, travellers can create a more sustainable city for everyone. Tourists can also support projects that work to green-up a city like cycling networks, kayaking tours, and conservation activities in city parks.

STOP AND THINK:
NOT JUST FOR HOLIDAY

Regardless of where we go or what we do, one of the main things that sets a holiday aside is our mindset. Most of us drop the tedium of everyday routine and open our senses more so than at any other point in the year. This is, of course, understandable, but imagine if we could channel this mindset on home turf? It's a notion that feels particularly relevant for cities because they depend on collective action. When I go away, I scrutinize the most sustainable options – where to stay and how to get around. I will quite happily explore Berlin by bike, book a sustainable architecture walk around New York, take tours around Rome with a migrant, and spend money in independent eco boutiques in Lisbon. I'll also give each activity 100 per cent of my mind and attention; I might even become impassioned and a little changed by what I find out. But how often do I do that in London, my home, where those convictions and insights would help me really make a difference? Hardly ever. While travelling to improve other places is important, making sure we do the same when we return home is essential.

HOTELS

Hotels can play a leading role in greening cities, too. Pioneering green hotel designs include the Radisson Blu Hotel in Frankfurt, which generates its electricity from the heat generated from the building, and Vo Trong Nghia Hotel's secret, cascading garden in Vietnam's Hoi An. In Luise, Germany, Creativhotel has created the world's first 'renewable hotel rooms' based on cradle-to-cradle sourcing. Everything is recyclable or biodegradable; the ceilings are made from straws and the carpet from recycled fishing nets.

In some cases, hotels are best placed to innovate green ideas that will hopefully catch on. The UAE-based developer Seven Tides is planning to use a hydroponic greenhouse on the World Islands to supply its portfolio of three hotels in Dubai. Hydroponics is a more efficient growing technique and can reduce water consumption by up to 90 per cent. Growing produce locally also reduces carbon emissions.

LJUBLJANA, SOVENIA

Europe's 'greenest city' knows this better than most. The quaint cobbled capital of Slovenia, Ljubljana, has won much acclaim thanks to its environmentally sensitive and long-term mindset. Between 1996 and 2006 the whole city went through a ten-year transformation to remove cars from the city centre. Planning instead focused on public transport and pedestrian and cycling networks.

While putting people (rather than cars) back into the heart of the city centre, the city also revitalized nature by planting over 2,000 trees, building five new parks, and making the banks of the River Sava more wildlife-friendly. Visitors have lapped it all up ever since, from whizzing between sites on electric Kavalir (Gentle Helper) buggies or strolling along the 7,000 tree-lined Path of Memories and Comradeship.

The transformation resonates with residents and visitors alike – thanks to car-free streets, the city is more 'authentic' than ever. It's also easier to get about, and the air is cleaner. Ljubljana's sustainable ethos has helped to bring locals and tourists closer; it's a far cry from the clashes that now dominate so many other European cities. Let's hope it stays that way.

TRANSPORT

Other cities that top the green charts include Copenhagen, Oslo, Gothenburg, and Amsterdam. Copenhagen has a Cycle Score for how bike-friendly each neighbourhood is, and Oslo is reopening its waterways to restore habitats and make them more accessible. Almost a third of vehicles sold in the city are now electric, too, and Gothenberg will soon get an electric river ferry. In the United States, Portland welcomes eco-minded visitors with over 180 miles of bike lanes, eighty bus routes and just under 100 train stations; unlike so many other cities on the West Coast, cars are less of a focus.

Some of the most successful green schemes have caught on worldwide. Paris' Velib was the world's first bike-sharing scheme, attracting 20 million users in its first year. The motivation was to cut dangerously high emissions in the city centre, but it soon turned out to be a novel way for tourists to get about, too. Thanks to its popularity, schemes cropped up all over from Shanghai to Madrid. In Berlin, for example – a vast, unwalkable city – it's possible to spend a weekend pedalling around rather than burdening transport or clogging the streets with taxis.

Campaigning for change

Cities are hotbeds of social and environmental innovation. It's here, among a crush of humanity and agendas, that activism and campaigns for a more equitable and sustainable future flourish. Our travels can encourage change. Socially minded tours are an excellent place to start. These could be led by marginalized groups, who benefit from the income but also help to breakdown misconceptions, or activists, eager to get their environmental or social message heard. Take a look at green travel guides like Conscious City Guides to find green fashion or food tours, and Urban Adventures has an In Focus Program working with non-profits. Better still, look for local listings like Social Enterprise UK, which can shine a light on the businesses putting purpose before profit throughout UK cities.

CASE STUDY

HARA HOUSE, INDIA

An inspiringly youthful example is Hara World in India. Friends Manoj and Jazzmine met while working for a not-for-profit in rural Rajasthan. After three years of debates about how best to channel their passion for youth empowerment and the environment, they established India's first zero-waste lodge, Hara House, in the city of Bikaner. From here, the Hara Revolution has blossomed into Hara World – an organization on a 'mission to use the power of travel to empower youth to step into their role as a leader for creating the world we want to live in'.

What makes Hara House unique is that it targets visitors and residents alike. Sifting through their programmes, from day tours of Mumbai's social enterprises to female empowerment courses, eco fashion collaborations, repair workshops, and leadership retreats, there's no 'us' and 'them'.

For Manoj and Jazzmine, creating cities that the youth population want to live in (environmentally and socially responsible) is nothing to do with tourism; it requires collective action.

CASE STUDY

CITY FARMS AND CLEANING PLASTIC, EUROPE AND US

Even city farms are getting greener. In 2020, Paris will reveal the world's largest urban farm – a 14,000-m^2 site on the top of an exhibition centre that will produce 1,000 kg of fruit and veg in high season. In New York, visitors can get back to nature on a trip to Brooklyn Grange Rooftop Farms, where tasting menus and workshops revolve around sustainable living and ecology. In Rotterdam, hotels and restaurants feed cows food waste scraps and collect milk from the world's first floating farm.

Travellers can also help to clean up cities. Rather than a chore, this can often provide a fresh perspective. In London, Copenhagen and Amsterdam, various plastic clean-up schemes get people retrieving rubbish from canals and rivers while taking part in kayaking, stand-up paddleboarding or boating. Guests on Copenhagen's GreenKayak can 'pay' for kayak rental by bringing back a bucket of rubbish gathered from the harbour and the canals. Founded in 2017, the not-for-profit and its volunteers have collected a total of 24,319 kg of plastic to date.

MIGRANTOUR, EUROPE

Throughout Europe, Migrantour demonstrates how migrants have shaped some of the continent's most famous cities. The organization describes itself as 'a form of responsible tourism at kilometre zero, which sees as protagonists fellow citizens also coming from distant worlds'. On a tour of Turin, Italy, a guide might take guests through Eastern Europe, South East Asia and the Middle East within the space of an hour. We so often celebrate cities as melting-pots of cultures, languages, beliefs and food from all over the world, yet how often do we absorb all those different perspectives when travelling around them?

TALAYSAY TOURS, CANADA

Tours led by refugees, indigenous groups, or migrants can also help to break down barriers. In Canada, First Nation owned Talaysay Tours guides visitors around Vancouver, the Salish Sea, and Stanley Park through the eyes of the Shíshálh (Sechelt) and Sḵwx̱wú7mesh (Squamish) First Nation. The dialogue shines a light on the city's relationship with nature and First Nation culture, both harmonious and destructive.

UNSEEN TOURS, UK

Tourism can also play a role in challenging preconceptions and uplifting the lives of marginalized people. In London, Unseen Tours invites tourists and locals to 'experience London's secrets through the eyes and voice of the homeless'. Twenty former homeless people lead tours around their patch – Pete reveals the eclectic back streets of Brick Lane (below), and Viv sheds new light on storied Covent Garden. The concept not only breaks down barriers and misconceptions around homelessness, but 60 per cent of ticket revenue is paid directly to the guides, and any additional profit is reinvested into the business.

STOP AND THINK: SLUM TOURISM

With one in eight people living in informal urban settlements, slums are a reality for cities all over the world. Therefore, it's no surprise that travellers and tourists are often intrigued. Slum tours come in all shapes and sizes – some, like Smokey Tours in Manila, are designed so that people living in the most impoverished conditions can tell their story. Others, like Reality Tours in India, hope to show entrepreneurial spirit and hope. But, however they are dressed up, are these tours primarily a means for wealthy visitors to gawp at some of the world's least fortunate people? The answer often lies in the visitor's intention, which makes it hard to pinpoint a right or wrong approach to slum tourism. Some people let respect lead their behaviour – taking the 'how would I feel if the tables were turned' approach, others see it as no different from going to the zoo. Once we've established that our intention is coming from a place of responsibility and respect, we need to scrutinize what tours we choose to support. Look for those that remove any 'us and them' barriers (like AWOL Tours in South Africa, which uses bikes rather than a minibus). Or, take an 'active' approach (Penda matches visitors with charities in South Africa's townships to provide professional photos for marketing materials). Favour small groups, locally owned and run, and tours where at least 80 per cent of revenue goes to people living in the informal settlements. Above all, remember we're all humans and deserve equal respect and rights.

Britz

Make it count

To justify travel in a climate and biodiversity crisis, we need to make it count. We've already explored how this might involve uplifting communities, empowering the marginalized, or protecting an ecosystem, so now it's time to get planning.

Each continent, country and place has its own set of needs, and if we want to travel sustainably, we need to let these influence how we travel. Rather than wander the world taking from it, this mini guide explores how we can give back.

For this chapter, I've been around the world and back to hunt down the most sustainable places to stay, experiences and ways to get around. The good news is, there were hundreds to choose from; here are a few for starters.

Oceania

Shrouded in ancestral legends and defined by landscapes of epic proportions, from wind-battered Pacific islands to Australia's barren interior, Oceania is as formidable as it is irresistible. The second least populated continent after Antarctica, Oceania holds endless opportunity for exploration: dusty road trips, marine hotspots, untouched coastlines, and scenic cities.

Around 70,000 years ago, people that were later known as Aboriginals were the first to migrate out of Africa. Over time, Papua New Guinea, the Pacific islands, Australia and New Zealand became home to hundreds of indigenous groups. These peoples subsided off the land, living in relative harmony and peace until European settlers trickled in from the 1600s. Throughout the next 200 years, religion, farming and industry changed the face of Oceania forever.

Indigenous peoples have suffered much during this time; often marginalized and forgotten. Although rights and respect are improving, there's still a long way to go. Responsible tourism helps visitors to support and learn from indigenous peoples, and it's an increasingly important tool in shifting mindsets, providing income and restoring pride. Travel is having another positive impact on Oceania, too. It's financing conservation to turn back the tide of destructive farming methods. In a continent increasingly at risk from extreme weather, restoring some of nature's balance is more pressing than ever.

Where to stay

KOOLJAMAN, KIMBERLEY, AUSTRALIA

On the tip of Kimberley's Dampier Peninsula, Kooljaman wilderness retreat lives up to its name. Accessible only by a rough track, the whole camp is off-grid and solar-powered. Run and owned by the Bardi Jawi people, nature and indigenous culture steal the show – every experience and moment is fused with a respect for both. ***[1]***

–

If you'd rather luxe it up, Sal Salis Ningaloo Reef is an off-grid eco-camp on an isolated stretch of coast in the Cape Range National Park.

ARKABA, FLINDERS RANGES, AUSTRALIA

Owned and run by conservation-minded Wild Bush Luxury, Arkaba is a sensitively converted homestead amid a 60,000-acre conservancy in the Flinders Rangers. Visitors can take leisurely safaris or embark on walking trails with camp-out stations and beds under the stars. Thanks to pest eradication, rare species like rock wallabies and western quolls are often spotted. ***[2]***

–

Wilpena Pound Resort is a more affordable option nearby, owned and operated by the Abnyamnthanha people.

HUON BUSH RETREATS, TASMANIA, AUSTRALIA

Deep in the Mount Misery Habitat Reserve – a conservation collaboration between several landowners – is Huon Bush Retreat's solar-powered, eco-minded cabins and low-impact campsite. Just 45 minutes from Hobart, the Huon Valley feels miles away, partly thanks to the 4-km gravel driveway that picks its way through untamed forest. ***[3]***

–

For a more extravagant experience, award-winning Saffire-Freycinet ticks all the boxes for eco-ultra-luxury on the edge of the Freycinet National Park.

THE BRANDO,
TETIAROA PRIVATE ISLAND, TAHITI

Brainchild of Marlon Brando while he was filming *Mutiny on the Bounty*, The Brando is one of the world's most sustainably minded luxury resorts. The entire island is LEED Platinum certified; solar panels create 70 per cent of the resort's energy, which also has the world's first Sea Water Air Conditioning unit. The Tetiaroa Society protects local wildlife and heritage. ***[4]***

–

Surrounded by similarly irresistible waters, the Jean-Michel Cousteau Resort in Fiji was one of the world's first eco-luxury properties.

ECO VILLA, CHRISTCHURCH, NEW ZEALAND

Exuding Christchurch's ancient connection to the land (the Māori name Ōtautahi means 'first place of plenty'), Eco Villa feels more like a country idyll than city centre abode. The sustainably minded hotel is home to an edible garden, outdoor bathtubs and upcycled furniture. The building was carefully put back together after the 2014 earthquake.

–

In Melbourne, Habitat HQ is an eco-hostel that uses 40 per cent renewable energy and encourages guests to get involved in local conservation programmes.

1

2

3

4

Experiences

DAINTREE RAINFOREST,
QUEENSLAND, AUSTRALIA

World Heritage-listed Daintree Rainforest is the oldest tropical rainforest in the world. It's also home to Cape Tribulation, where rainforest meets the Great Barrier Reef. Responsible tourism helps to fund on-going conservation efforts. Experiences are low-impact and mindful of heritage, including electric boat river cruises and award-winning Dreamtime Walks. *[1]*

VOYAGER ESTATE, MARGARET RIVER,
AUSTRALIA

Forty-year-old, family-owned Voyager Estate is one of Australia's most established wineries, and since 2018 it's been making the transition to becoming 100 per cent certified organic by 2023. The estate's responsible approach to wine making and organic principles are woven throughout all visitor experiences. Over 70,000 native trees have been planted, too.

WUKALINA WALK, TASMANIA, AUSTRALIA

Australasia is awash with Aboriginal owned and operated walks and experiences; on Tasmania, one of the most renowned is a multi-day guided Wukalina hike around the Bay of Fires wilderness. The tour offers the chance to meet Palawa elders – the only group of humans to evolve in isolation for 10,000 years.

WHALE WATCH KAIKŌURA, NEW ZEALAND

Thanks to a 3-km-deep canyon along its wild coast, Kaikōura is one of the few places in the world where sperm whales congregate year-round; the most sustainable way to catch a glimpse is on a Whale Watch Kaikōura tour. Māori owned, WWK provides jobs for the Ngati Kuri community and has secured Māori ownership of the Kaikoura Peninsula. *[2]*

KOLOMBANGARA, THE SOLOMAN ISLANDS

A perfect cone rising out of the sea, Kolombarangara's dormant volcano and biodiversity-rich slopes create a quintessential South Pacific Island. Although fiercely protected by the indigenous Dughore people, logging is rampant. By taking a hike with the local conservation association (organized by Oceania Expeditions), you can help to keep what remains intact.

3

4

Getting Around

BRITZ ELECTRIC CAMPERS, NEW ZEALAND

Kiwi adventure company Britz have launched New Zealand's first electric campers, seriously upping the sustainability credentials of one the world's most epic road trips. Available for hire out of Auckland and Queenstown, the eVolve has a driving range of 120 km and Britz covers the cost at selected charging points all over the country.

JOURNEYS BEYOND RAIL, AUSTRALIA

The Indian Pacific line links Sydney with Perth in just three days, passing through Adelaide and the other-worldly Nullarbor Plain on route. Equally romantic is the Ghan, named after the Afghan camels who used to carry supplies up to Alice Springs before the railway was built. It takes 48 hours from Adelaide to Alice Springs, and another 24 hours to Darwin. *[3]*

CANOEING KATHERINE RIVER, AUSTRALIA

In Australia's Northern Territory the towering cliffs of Nitmiluk Gorge plunge into Katherine River. Eagles fly overhead, freshwater crocs roam the banks, and black cockatoos look on inquisitively. From lush pandanus to paperbark forest, the landscape is best experienced via multiday canoe trips; for a top eco experience, head out with Gecko Canoeing. *[4]*

CORAL EXPEDITIONS, OCEANIA

These Australian Eco-accredited small (120 passengers maximum) expedition ships are about as sustainable as a cruise will ever get. Produce is local and seasonal, itineraries are designed to ensure communities benefit, unique experiences take priority over extravagance, and guests are educated about conservation and heritage issues along the way.

TRADITIONAL SAILING, SOUTH PACIFIC ISLANDS

The Okeanos Foundation hasn't been designed for tourists, but that's what makes it so appealing. Its mission is to help empower Pacific Island people via traditional, sustainable sea transport. The sailing canoes use green energy – wind, sun and coconut biofuel. To help fund the mission, Oceania Expeditions can arrange one to three-day itineraries.

North America and the Caribbean

Idealized, satirized, and worshipped on the big screen for decades, North America is no stranger to the world. Yet first time visitors never fail to be bowled over by its sheer size. From New York's high-rise living to the Arctic's resourceful Nunavut communities and all the mountains, deserts, cities and cultures in between, there are no limits to adventures.

Often defined by its headstrong global voice, it can be easy to forget that North America is largely beyond human control. Between the ancient Appalachian Mountains and the considerably taller Cordilleras, the weather-savaged Great Plains dip into vulnerable deltas and low-lying peninsulas. This is a continent of extremes, and is unique in covering every single climatic zone. Since the first European's settled in America, humans have manipulated the landscape to meet their needs, which now exceed the rest of the world. The continent has the highest average income and food intake per person, and consumes more than four times the world's average energy.

Travel can play a part in turning back the clock. The right experiences champion the rights of indigenous groups, and prove the value of conservation over intensive, large-scale farming. Travellers can also play a role in encouraging this progress-hungry continent to channel innovation for good, by investing in green technology and sustainable solutions.

Cascading down into the balmy Caribbean Sea, responsible travel throughout the Carribean's 7,000 islands looks beyond the resorts and cruise ships to support local businesses, protect fragile ecosystems, and take time to delve into often overlooked heritage. Initiatives like the new sustainability collaborative See the Caribbean are proving that there's so much more to these islands than lying around on a beach.

1
HOTEL

2

Where to stay

FLAT CREEK RANCH, WYOMING, US

Wyoming is one of the most sustainable destinations in the United States thanks to 98 per cent of land being under some form of protection. There are numerous dude ranches and lodges to help visitors make the most of it, but one of the most sustainable is Flat Creek. Here, operations from recycling to energy use are continually scrutinized and improved.

–

Conservation-led ranch experiences can also be found at Ted Turner's mesmerizing series of lodges in New Mexico.

1 HOTELS, VARIOUS LOCATIONS, US

W Hotels founder Barry Sternlicht's eco-chic hotel chain is a good bet for those looking for mindful, boutique digs in New York, Hollywood, and Miami. Each property meets the US Green Building Council's LEED certification, and most are plastic-free front of house, use reclaimed materials, and prioritize biophilic design.

–

Eaton DC is another slick city hotel that punches above its weight when it comes to using hospitality as a force for good via social empowerment and eco education.

THE GLADSTONE HOTEL, TORONTO, CANADA

Canada's only hotel with a B Corp certification (cross-industry sustainability certification) is also Toronto's oldest – it's been in operation since 1889 and has been recently refurbed with sustainability top of mind. Green roofs, non-toxic cleaning products, a strict waste policy, and championing local social justice groups, are just a few of the eco creds to note. ***[1]***

–

On the other side of Canada, Skwachays Lodge in Vancouver is an award-winning social enterprise funding indigenous housing and arts.

FOGO ISLAND INN, NEWFOUNDLAND, CANADA

Beaming into Newfoundland's constantly shifting skies and floating above a wild and bleak granite perch, Fogo Island Inn's contemporary design is a sustainability icon. It's owner, Zita Cobb, pioneered the Economic Nutrition^{CM} mark, which informs guests exactly where their money goes. In Fogo Island's case, a 12 per cent surplus is reinvested in community projects.

–

For equally breath-taking ocean views without the price tag head over to the West Coast to stay in the 115-ft-tall Pigeon Point Lighthouse hostel, an hour south of San Francisco.

CASTARA RETREATS, TOBAGO, TRINIDAD AND TOBAGO

Forget the overdone extravagance of so many Caribbean resorts. Castara Retreats, overlooking one of Tobago's prettiest bays, exudes warmth and spirit far more in keeping with the location. Staff are invited to become stakeholders, the resort blends seamlessly with the surrounding village, and everything is led by locals.

–

A little further north, Jungle Bay in Dominica is a largely solar powered resort built among 55 acres of rainforest. ***[2]***

EL PATIO 77, MEXICO CITY, MEXICO

History, art and sustainability blend seamlessly in this refurbed nineteenth-century townhouse that hopes to share 'a new future for Mexico City's tourism'. Despite the old-world feel, the B&B has the latest green tech – solar water heaters, rainwater harvesting and filtered water. Artefacts and handicrafts throughout the hotel have been sourced directly from artists.

–

For another city bolt-hole with heart try Home in Buenos Aires that has a strict local sourcing and waste sorting policy.

1

2

Experiences

ALCATRAZ CRUISES, SAN FRANCISCO, US

Alcatraz's cruise ferry boats are some of America's greenest. Toxic emissions are 80 per cent lower than California's strict Tier 2 air quality requirements, thanks in part to electric motors, wind turbines and solar panels to top up clean fuel. Crew are trained to sort and recycle all waste, too. *[1]*

AMERICAN PRAIRIE RESERVE, MONTANA, US

Spearheading a unique effort to create the largest nature reserve in the United States by linking together more than 3.5 million acres of critical habitat, this is one of the world's most critical grassland conservation projects. On the legendary plains of Montana, visitors can sleep under the stars, hike, or take part in wildlife conservation efforts. *[2]*

BROOKLYN GRANGE FARM VISITS,
NEW YORK, US

Green roof and rooftop soil farm experts Brooklyn Grange offer visitors a slice of country life overlooking Manhattan's skyline form their Northern Boulevard base. Alongside growing produce (the three New York City farms produce an incredible 80,000 lb of organic goods a year), events, tastings and workshops demonstrate why greening urban spaces is a win-win.

KUALOA RANCH, KANEOHE, HAWAII, US

In windward O'ahu, Kualoa Ranch is home to Jurassic Park and conservation fame. Kualoa's vision has always been to enrich people's lives by preserving the land and its heritage; day excursions and cultural programmes are designed in harmony with nature and seek to shine a light on Hawaii's ancient culture.

INDIGENOUS ARTISAN'S TRAIL,
OKANAGAN, CANADA

The award-winning destination, Thompson Okanagan Tourism Association, will soon launch an Indigenous Artisans Trail. A mobile trailer filled with art from the Interior Salish Indigenous Peoples – Secwépemc, Nlaka'pamux and Syilx – will roam the region. The hope is that it will provide income and bridge any gaps between visitors and these groups.

Getting Around

EAST COAST GREENWAY, US

The United States has plenty of renowned long-distance cycling routes but one of the most exciting is the new East Coast Greenway. Though currently only 23 per cent complete, the route will link Calais, Maine, and Key West, Florida, with a 3,000-mile car-free cycling trail. If you can't wait, at the moment, the remaining 77 per cent is on low-traffic roads. *[3]*

PARK BUS, CANADA

North America's national parks can suffer from overtourism so visitors need to be mindful of when and how to travel around them. One novel service is Park Bus that operates coaches between cities and wilderness areas throughout Canada. Yosemite National Park has also introduced a park and ride electric bus service to reduce congestion.

ROCKY MOUNTAINEER AND AMTRAK'S COAST STARLIGHT, CANADA AND US

North America isn't renowned for the most comfortable rail travel but two of the best options are Canada's Rocky Mountaineer between Vancouver and Jasper, and the United States' west coast line, which is considerably smoother than Amtrak's cross-country routes. *[4]*

JUS' SAIL, ST LUCIA, CARIBBEAN

Protecting heritage, saving carbon and uplifting communities is all rolled into one with this innovative Caribbean yacht charter company. Operating a carefully restored traditional Carriacou Sloop called Good Expectation, Jus' Sail provides visitors with the chance to explore St Lucia under wind power, and uses the funds to upskill keen local sailors.

COPPER CANYON TRAIN TOURS, MEXICO

South and Central America isn't renowned for its comprehensive rail network but there are pockets of scenic trips. One of them meanders from the coast to the Copper Canyon, which is four times larger than the Grand Canyon. Over four hours, on the first-class express, the train clatters almost 400-miles through Mexico's rugged interior to the Sierra Madre Range.

South and Central America

A continent of superlatives, South America is home to the world's driest desert, biggest rainforest, longest river and tallest waterfall. With such a variety of treasure to explore, few people visit once, and many consider its impenetrable landscapes addictive. Between the macaw-filled thickets of the Pantanal in Brazil and allusive pumas in the wilds of Patagonia, this is certainly no place to rush.

As intricate as its geography and ecosystems, is South America's culture. Peru was home to one of the world's first civilizations, the Norte Chico. Since then, independent peoples had lived a relatively separate existence until the Portuguese and Spanish colonization of the fifteenth century. Sandwiched between two warm seas, modern history has brought with it centuries of migration, conquest and trade, endowing the continent with a kaleidoscope of heritage. Although episodes of revolt and corruption have stunted some countries since, indigenous cultures here remain some of the world's strongest.

Alongside such unique human stories, South America is also the earth's lungs. The Amazon stores almost 200-billion tonnes of carbon, and 35 per cent of it is protected by indigenous peoples. In the midst of political turmoil, and climate-denying leaders, making the case for conservation is more urgent than ever. Tourism is stepping up to the challenge, and from jaguar habitation programmes to the world's largest rewilding initiative, there is hope. Travellers can encourage progress by supporting grassroots sustainable initiatives in urban and rural contexts.

Where to stay

POUSADA TRIJUNÇÃO, CERRADO, BRAZIL

Often overshadowed by the Amazon, the Cerrado tropical savannah ecoregion covers 20 per cent of Brazil and is one of Conservation International's thirty-five biological hotspots; without it, many other ecosystems in South America would collapse. This low-lying *pousada* (Portuguese for hotel) uses tourism to fund conservation, including monitoring the Maned Wolf and supporting wildlife-friendly farming for locals.

–

Equally vital conservation work is going on in Brazil's Pantanal, led by eco-tourism legend Roberto Klabin at Caiman Ecological Refuge. ***[1]***

TAHUAYO LODGE, IQUITOS, PERU

A comfortable, waterside lodge in the heart of the Amazon, Tahuayo was recently awarded Peru's highest National Environment Award for its efforts training local people to become guides. Visitors also learn how to explore the fragile environment with minimal impact. The lodge provides local employment, scholarships and supports research. *[2]*

–

Connected to the Mamirauá Sustainable Development Reserve, the floating Uakari Lodge in Brazil is community-run, sustainability-focused and conservation-minded.

ECO CAMP PATAGONIA, TORRES DEL PAINE, CHILE

This multi-award-winning, low-fi camp surrounded by Chile's most-prized mountains, rivers and forests, was one of the first places in the world to create geodesic domes for visitors. From kayaking to hiking, activities are as low-impact as the accommodation, which complies with the environmental management ISO14001 standard. *[3]*

–

Geodesic design also maximizes views while minimizing impact at Kachi Lodge in the Bolivian Uyuni Salt Flats.

JICARO LODGE, GRANADA ISLETAS, NICARAGUA

A sun-drenched, jungle-filled outcrop surrounded by Lake Nicaragua's serene waters, Jicaro Lodge is the ultimate private island retreat. Nine simple and elegant casitas were each built using reclaimed timber from Hurricane Felix. The property is managed by sustainability pros, the Cayuga Collection. *[4]*

–

Down in the 'motherland' of sustainable travel – Costa Rica – Lapa Rios and Pacuare Lodge are run on similar principles.

1

2

3

4

1

2

Experiences

PACHAMAMA JOURNEYS, ECUADOR

Run by the Pachamama Alliance, which empowers indigenous peoples to protect their land and inspire others to lead more sustainable lives, Pachamama Journeys take place in the Amazon or High Andes of Ecuador. The small-scale, ethical trips immerse travellers in ancient cultures and pristine landscapes. *[1]*

JAGUAR SPOTTING, BRAZIL

Deep in the world's largest wetland, The Pantanal, Caiman Ecological Refuge works with non-profit Onçafari to help protect one of the world's most allusive cats: the jaguar. Through a careful habituation (making the animals used to safari vehicles) and local conservation education programme, lucky visitors get the chance to experience jaguars in the wild. *[2]*

COMMUNITY-LED ADVENTURES, GUYANA

Striving to become the world's most sustainable destination, Guyana has developed a fiercely community-led and eco travel network. One of its newest ecolodges, Warapoka, offers unique catch-and-release fishing opportunities. Day trips to the Amerindian village of Moraikobai include cassava flour making and bird spotting with the community.

EAT LIKE A LOCAL, MEXICO CITY, MEXICO

Run by Rocio, who was born and bred in Mexico City, Eat Like a Local's 'Food Safaris' take visitors and their taste buds on a whirlwind tour of Mexico City's lesser-touristed street food stalls, markets, and family-run restaurants. Responsible to the core, 100 per cent of profits stay in Mexico City.

GRAFFITIMUNDO, BUENOS AIRES, ARGENTINA

Designed to help visitors get a deeper understanding of Buenos Aires' politicized street-art movement, Graffitimundo is a not-for-profit that works in close collaboration with artists. Profits from the tours help provide up-and-coming or disadvantaged artists with workshop and gallery space to create and sell their work.

Getting Around

CAMINO DE COSTA RICA, COSTA RICA

This non-profit association has created a walking route to ensure that tourism reaches rural areas that need the income most. The consequent trail is a 280-km and sixteen-day romp across the whole country taking in volcanoes, cloud forests, and rolling hills. Overnight stays are mostly in hostels, homestays and family-run lodges.

CARRETERA AUSTRAL, CHILE

Climbing, circling, and descending Chilean Patagonia's mountain landscapes for 810 miles, the Carretera Austral is a biking route from Puerto Montt to northern Patagonia. It crosses two national parks, and the indigenous-owned forests surrounding the Puyuhuapi Hot Strings. Ferries join up some sections, giving the legs a welcome break.

LONG-DISTANCE BUS, SOUTH AND CENTRAL AMERICA

The pinnacle of long-distance bus travel has to be in South and Central America. The experience peaks in Argentina, where slightly pricier top-deck seats recline to an almost horizontal position and come with the odd free drink and snack. Distances are huge and journeys can take days but there's no better way to understand the scale of the continent.

RIVERBOAT FROM BELÉM TO MANAUS, BRAZIL

Public riverboats are the most affordable and sustainable way to cruise down the world's longest river: the Amazon. The trip takes six days and guests can opt for a simple cabin or hammock space on the deck. Going slow provides plenty of opportunities to catch a glimpse at pink river dolphins, and chat into the night with fellow passengers.

MARIO NELSON, GUIDE FOR JAGUAR-CONSERVATION GROUP ONÇAFARI, CAIMAN ECOLOGICAL REFUGE AND LODGE, BRAZIL

Growing up on a cattle farm in the Pantanal, Mario comes from a family of former hunters that he has been instrumental in converting to conservation.

–

How has eco-tourism impacted your life?

I've been guiding with Onçafari at Caiman for four years and have learned so much through the experience and the training. It's changed the way I am. I now love nature and being in the wild. It gives me great satisfaction to be able to show people my culture and the beauties of the Pantanal.

What effect has it had on your family and friends?

It was a huge challenge for me in the beginning. My family comes from a different background – one based on cattle ranching and taking care of these animals, and they did not have a liking for conservation or for eco-tourism. But I managed to change their views on those subjects, showing them that through eco-tourism people in the region have better job opportunities.

Do you think tourism has a positive or negative impact in your area?

Definitely a positive impact because apart from protecting the animals, it brings better job opportunities to the region and more income for many families.

What are your hopes for the future?

That more people get the opportunity to work conserving wildlife, so they become motivated to protect more land. Personally, to be a bilingual guide, to travel and to get to know the world better.

Europe

Between the birthplace of democracy in Athens, the East-meets-West gateway of Istanbul, and London's layered architectural canvas, most travellers flock to Europe for historical intrigue. It's hardly surprising given that thousands of years of history is crammed into one of the world's most populated landmasses. Nowhere else is it possible to witness such a myriad of beliefs and human endeavour in such short distances.

Despite the urban sprawl and crowded cities, Europe's got its fair share of wilderness and geographical interest, too. Often referred to as 'the peninsula of peninsulas', thanks to its many tendrils carving into and around seas and oceans, intricate coastlines create unique ecosystems. The consequent dramatic landscapes can come as a surprise, and are largely unknown even to European residents.

From Romania's expansive Carpathian Mountains to Greece's snow-capped spine and the plummeting gorges of Spain's Pico's de Europa, getting off the beaten track is well worth it. It's important, too, given that overtourism is so prevalent. Travellers can make their trip count by supporting the growing rewilding movement; many schemes and initiatives would not exist without visitor income. Although Europeans are some of the most environmentally minded in the world, action varies, so it's important for visitors to watch their footprint. The same goes for social equality and inclusiveness.

1

2

Where to stay

KORLARBYN ECOLODGE, VÄSTMANLAND, SWEDEN

Deep in Bergslagen Forest, just two hours from Stockholm but home to utter silence, moose and beavers, Korlarbyn offers the ultimate off-grid, low-fi experience. Huts, nicknamed 'hobbit houses', are converted hunting shelters, and there's no shower or electricity. Evening meals and campfires tend to be communal. *[1]*

–

If you prefer to glamp with big views and a little more comfort, White Pod Hotel in Switzerland is the original European eco geo-dome experience.

GOOD HOTEL, LONDON, UK

A former prison that floated over to London's Victoria Dock from Amsterdam, Good Hotel has plenty of intrigue going for it. But it's the hotel's responsible ethos that turns most heads. Founder Marten Dresen operates the hotel according to his vision of 'social business'; profits are reinvested in social causes, and the hotel supports the local long-term unemployed.

–

With the motto 'stay open-minded' Magdas Hotel, Vienna is another champion of social business, employing refugees from over fourteen different countries.

THE SCARLET HOTEL, CORNWALL, UK

A pioneer of sustainable luxury, this beachfront hotel surrounded by natural pools, meadow gardens, and the work of local sculptors, has been ticking all the sustainable boxes long before anyone else got on board. The hotel's list of 111 sustainability creds include the careful rehousing of reptiles pre-construction and slippers made from recycled bottles. *[2]*

–

If you prefer eco beach digs with a little more privacy, head to Eco Casa Penna in Sicily for paired-back mod-cons run on solar power.

MONTAGNE ALTERNATIVE, SWITZERLAND

When co-owner, Benoit Greindl, discovered the village of Commeire high in the Swiss Alps, only twelve residents remained. The rest had left because agricultural income had dried up. Since then, he's worked with local craftsman to transform village buildings into a nature-led retreat and hotel, breathing life back into the abandoned mountains.

–

A similar project in a very different setting is Milia in Crete, where fifteenth-century cottages and apartments have been brought sustainably back to life via tourism.

I PINI, TUSCANY, ITALY

Proving that going vegan and organic doesn't mean sacrificing on wine, flavours and views, Agrivilla i Pini is a delicately converted Italian mansion surrounded by Tuscany's rolling hills of pines, figs and vineyards. The onsite organic garden is grown on permaculture principles, and the 5-hectare vineyard is deliciously chemical-free

–

The UK's first vegan hotel, Saorsa 1875, opened in Scotland in 2019 offering a 'stay in style without compromising ethics'.

1

2

Experiences

CLEAN A BEACH, UK AND EUROPE

Surfers Against Sewage (SAS) is the UK's largest plastic-free campaign group; so far it's removed over 250,000 kg of plastic pollution from rivers, seas, streets and parks. Several times a year, SAS organizes a Europe-wide beach-cleaning weekend and encourages everyone to get stuck in. For other beach cleans, check out Surfrider Foundation's Ocean Initiative. *[1]*

THE BIG PICTURE, THE CAIRNGORMS, SCOTLAND

Thanks to rising awareness of the importance of rewilding, there are heaps of conservation experiences up for grabs all over Europe. These include trips to track bison in Poland with The European Safari Company and protect bears in Italy with The European Nature Trust. Rewilding advocacy group The Big Picture runs week and day tours in Scotland. *[2]*

KOKS, FAROE ISLANDS

Often lauded as the world's most remote restaurant, KOKS might also be Europe's most sustainable Michelin-star. From the simple abode tucked between the island's austere mountains, Head chef Poul Andrias Ziska 'distils the taste and smell of the Faroese landscape' using fiercely local ingredients and traditional techniques.

JAZZ NIGHT EXPRESS, ROTTERDAM TO BERLIN, THE NETHERLANDS AND GERMANY

Dismayed that night trains no longer run from Amsterdam, since 2016 Noord West Express has put on its own 24-hour sleeper service from Rotterdam to Berlin via Amsterdam and back. The benefits of running your own service means you can fill it with plenty of fun – including jazz bands, DJs, and a 'gastronomic journey' in the dining cart.

ALPINE PEARLS, THE ALPS, SWITZERLAND, AUSTRIA AND FRANCE

Proving that holidays across Europe's most spectacular mountains can be car-free, Alpine Pearls joins together twenty-one Alpine villages via train and bus. The villages – from South Tyrol in the Dolomites to Chamois at the foot of the Matterhorn – have been chosen for winter and summer activities, and sustainable places to stay.

3

4

Getting Around

PICK UP A PILGRIMAGE, EUROPE

Old ways and pilgrimage routes crisscross all over Europe's mountains, hills, cities and coasts. Explore Roman ruins, beaches and waterfalls on the Lycian Way or cathedrals and chalky downs on the British Pilgrimage Trusts' Old Way. The most famous is the Camino de Santiago across northern Spain, but it pays to look for less crowded routes. *[3]*

FAIR FERRY, ROTTERDAM, THE NETHERLANDS

Most people know that ferries aren't the most carbon efficient way to get around, but a few innovative start-ups are changing that. One is Fair Ferry, which opts for wind power over fuel. It offers guests long or express sailing trips from Rotterdam to London, and even across the Atlantic.

AVENUE VERTE, FRANCE AND UK

This Anglo-French initiative links London and Paris with 247-miles of traffic-free or back road cycling. One of the most pleasant sections is an asphalt disused railway line that meets cyclists off the Channel ferry in Dieppe. Meandering through orchards and passing dilapidated old country mansions is a pleasant way to discover northern France.

THE ROVER PASS, SCOTLAND

Rail connections in Europe are some of the best in the world, and improving all of the time. A great new affordable option in Scotland is the Rover Pass, providing up to eight days of unlimited rail (and ferry) trips all over the country from Glasgow to John O'Groats. For more ideas, check out Green Traveller's flight-free guides. *[4]*

ELECTRIC ROAD TRIP, SWITZERLAND

Europe is well-equipped for an electric road trip, and new itineraries are popping up all the time. One of the latest is spearheaded by My Switzerland; the 'grand electric road trip' carves through almost 2,000 km of scenery, including twenty-two lakes, Alpine passes, vineyards, and chocolate-box villages.

Africa

Perhaps the world's most renowned wildlife destination, there is so much more to Africa than the Big Five. Alongside epic migrations and gorilla-filled forests, there are mountains to climb, deserts to get lost in, and seas to explore. From Uganda's unique Ruwenzori to the Gulf of Guinea's fragile mangroves and deltas, and Morocoo's High Atlas Mountains to the Kalahari's shifting sands.

Home to the world's first humans (*Homo sapiens* stomped on African soil at least 200,000 years ago), cultural interest is hardly amiss. From Ethiopia's rock-cut monolithic churches to Zanzibar's spice markets, heritage-seekers would take years to unravel Africa's intricate history. But for all its wildlife and culture, modern Africa demands more of our attention. In places only recently battered by corrupt leaders, civil wars, and colonialization, new shoots of hope, creativity, and innovation are staggering.

With one of the world's fastest-growing populations and rich in natural resources, landscapes are under-threat all over Africa. Tourism plays a huge role in supporting the work of conservationists. Whether saving a species from poachers or protecting an ecosystem from mining or farming, the most sustainable travel experiences revolve around people. Making travel count in Africa means empowering communities and improving economic prospects, rather than funding a new form of imperialism.

Where to stay

LATITUDE HOTELS, KAMPALA, UGANDA, LUSAKA, ZAMBIA AND LILONGWE, MALAWI

This new hotel chain has ambitious plans to change the face of African hospitality. It hopes to create a space for and by creatives in some of Africa's most vibrant cities, partly to lure visitors away from wildlife. Social impact and environmental care is fused into every decision.

–

Mid-range Acacia Tree Lodge in Nairobi puts all of its profits into a foundation funding education for kids from Nairobi's slums.

KYAMBURA GORGE LODGE, VOLCANOES SAFARIS, UGANDA

Volcanoes Safaris is a sustainably minded stalwart of Ugandan and Rwandan hospitality; it was the first operator to open gorilla camps, import nothing, and pay a fair wage. In 2019, the company broadened its conservation game via the Kyambura Gorge Eco-tourism project to safeguard the 'Valley of the Apes'. ***[1]***

–

With a similar community-focus, but on a much humbler scale, Three Tree Hill Lodge in South Africa is a family-friendly guesthouse with far-reaching Kwa-Zulu Natal views.

DUNE CAMP, WOLWEDANS, NAMIBIA

Wolwedans portfolio of luxury camps in the seemingly desolate south-west of Namibia have been instrumental in the creation of the NamibRand Nature Reserve; over 400,000 acres of undisturbed wilderness. Founding member of The Long Run, each low-impact camp is operated according to the 4Cs – conservation, community, culture and commerce. ***[2]***

–

One of Africa's oldest safari outfits, Cottar's Safaris, recently opened a tented Conservation Camp that helps to fund the surrounding community-owned conservancy.

MAGASHI CAMP, WILDERNESS SAFARIS, BOTSWANA

Besides from some hefty conservation work, contributing to the financial sustainability of Rwanda's Akagera National Park, Magashi has a light footprint. The camp is 100 per cent solar-powered and single-use plastic free. One-third of staff are from the surrounding communities, with more in training, and the Wilderness Eco-Club helps sixty kids get up-close with nature.

–

Camp Nomade is an equally eco-minded camp (it's only open part of the year) where nature steals the show in the lesser-visited Parc National Zakouma, Chad.

DOMWE ISLAND ADVENTURE CAMP, LAKE MALAWI, MALAWI

Back-to-basics doesn't get much more idyllic than this hideout on the shores of Lake Malawi. There's no electricity, so lighting is by solar, paraffin lamps and wind-up torches. Safari tents take little toll on the landscape and the toilets are dry composter. Activities include hiking, swimming, and kayaking.

–

Chumbe Island Lodge is one of Africa's most highly acclaimed conservation-minded lodges on a tiny island off the coast of Zanzibar.

1

2

1

2

Experiences

CONGO CONSERVATION COMPANY, DEMOCRATIC REPUBLIC OF CONGO

Perhaps the continent's least-explored wilderness, the Congo Basin is a dense and complex ecosystem home to lowland gorilla, forest elephant and forest buffalo. The Congo Conservation Company operates expeditions through the Odzala-Kokoua National Park and Sangha Trinational to conduct research and help prove nature's worth. *[1]*

SPORT WITH A PURPOSE, MALAWI

Ticking off giving back and getting healthy in one swift and epic swoop is Orbis Expedition's Sport with a Purpose. Trips include trail running with endurance runner Susie Chan, cycling with round-the-world cycling ecord holder Mark Beaumont and hiking with adventurer and author Sarah Outen. Also check out Singita's Grumeti Fund Run and the Somalia Marathon. *[2]*

VOLUNTEER WITH SEED, MADAGASCAR

By volunteering with sustainable development charity SEED Madagascar, travellers can leave a positive mark on this heavily de-forested island. Opportunities to get involved include healthcare, livelihood, conservation and education projects, and 92 per cent of volunteer fees go directly to the charity.

MAURITIUS CONSCIOUS, MAURITIUS AND LA REUNION

Better known for fly-and-flop resorts, Mauritius and neighbouring La Reunion are home to quaint guesthouses, adventure-ready jungles and mountains, and eclectic cultures. This eco tour operator helps visitors to get off the beach to enjoy locally owned and led experiences that revolve around nature and heritage.

TASTEMASTERS AFRICA, ACCRA, DAKAR, CAPE TOWN AND JOHANNESBURG

Described by Fusion as the 'sexy, dope way to experience Africa', Tastemakers is on a mission to disrupt the African travel narrative. Each experience – from hitting nightlife with DJs to art tours – is carefully selected by the team for authenticity. Local guides and curators are often passionate creatives or activists.

3

4

Getting Around

REMOTE N WILD WALKING SAFARIS, KENYA

These remote walking safaris are not only low-impact and led by local Samburu, but are operated by and for a conservation charity. The Milgis Trust protects 6,000 km^2 of precious landscapes to the West and North of Lenkiyou (Matthews) Range. It helped to reintroduce elephants to the area after the population was devastated by poaching.

SELINDA CANOE TRAILS, BOTSWANA

There is no better (or more sustainable) way to experience Botwana's life-filled channels than the 40-km, four-day paddle down the Selinda Trail. Carving slowly through the 320,000-acre Selinda Reserve, home to elephants, buffalo and African painted wild dogs, wildlife is abundant and viewing opportunities plentiful. *[3]*

TANZANIA BY TRAIN

Rovos Rail and the Blue Train offer Africa's most iconic train trips, but for those that want something a little more intrepid, you can't beat the good-value 1,860-km journey from Dar es Salaam to Kapiri Mposhi. First class sleeper cars, a restaurant car and lounge are available on board. Wildlife spotting as the train clips the Selous Game Reserve is a highlight.

EAST AFRICAN SAILING EXPEDITIONS

Sailing is a wonderful way to experience East Africa's vast and pristine coastline, and The Explorations Company has a series of trips on offer. An Arab dhow takes visitors through the Mozambiques Quiribas, camping each night on a beach, or around-the-island Zanzibar trips revolve around Swahili's sea-faring culture.

CYCLE AROUND RWANDA

Often referred to as the land of a thousand hills, Rwanda offers some of Africa's most demanding and rewarding cycling adventures. The most famous route is the 227-km Congo Nile Trail, which skirts Lake Kivu weaving through countryside and villages for five days. Swot up first at the Africa Rising Cycling Centre, home to the Rwandan cycling team. *[4]*

Middle East

So mesmerizing is the Middle East's cradle of civilization, that it's better known from legend than from history books. The concentration of ancient artefacts and civilizations expand well beyond the Pyramids; the whole of Egypt's Nile Valley is awash with history, as is Jordan's Petra, Iran's Persian Isfahan, and Lebanon's Roman ruins. It's here that the first cities were built, and three of the world's great religions materialized. Today, it's also the birthplace of ultra-luxury and high-rise cities.

Beyond human endeavour, there is wilderness a-plenty. The Middle East is home to storied rivers like the Nile and Euphrates and mighty deserts, including the Sahara, the Empty Quarter and Wadi Rum. Between Iran's peaks and Egypt's Red Sea, adventures by foot, jeep or boat intersect arid national parks, lush hills, snow-capped summits and even coral-filled seas. As compelling as the natural beauty, is the unrelenting Arabic hospitality. Whether strolling through the bazaars of Muscat or Cairo's narrows, it's not unusual for a fleeting conversation to continue over tea.

By choosing more sustainable and responsible travel experiences, visitors not only contribute to the Middle East's social and environmental progress, but unravel the truth beyond the stereotypes. In a region often associated with oil, civil war, poor human rights and large-scale developments, travellers can help to uplift more positive voices. When managed responsibly, tourism can protect the unique ecosystems, secure cultural heritage, and rebuild communities, too.

Where to stay

FENYAN ECOLODGE, JORDAN

This 100 per cent solar-powered lodge, which blends seamlessly with Wadi Araba's rock-strewn wilderness, has been championing eco hospitality for over fifteen years. The lodge was built by the Royal Society of Nature but is operated by Jordanian-owned EcoHotels. The premise has always been to uplift the Dana Biosphere Reserve's Bedouin communities via tourism.

–

In quite a different setting, IHG has opened Dubai's first solar-powered hotel, Hotel Indigo Dubai.

SOULY ECOLODGE, OMAN

In a continent awash with extravagant resorts, Souly's simple premises are refreshing. Perched on Salalah's wild coast, the series of bungalows were built using local and recycled wood and stone. Bathrooms are stocked with organic products, food is as local as possible, and experiences revolve around nature.

–

For a more luxed-up desert experience, Alila's Jabal Akhdar, Oman was built according to LEED principles and has a clever and traditional irrigation system to save water.

BKERZAY, LEBANON

Sculpted into the hills using local stone, the Bkerzay 'village' is a conservation project surrounded by olive groves and wooded hills just 45 minutes from busy Beirut. An on-site purification station recycles water, and solar energy helps power the guesthouses. Each guesthouse is modelled on traditional Lebanese design, and was built around existing trees.

–

Built using age-old building techniques, Naritee Ecolodge near Yazd, Iran also sensitively blends into its surrounds. **[1]**

AL MAHA, DUBAI, UAE

Just under an hour from the city centre, this luxury tented camp sprawls across an enviably sheltered spot in the heart of the Dubai Desert Conservation Reserve. Designed with Bedouin culture and conservation in mind, the camp was instrumental in Emirates Hotels & Resorts' decision to create the reserve, which covers 5 per cent of Dubai's total land area.

–

Another curious setting to spend the night is the new Al Faya Lodge, west of Dubai, which sits in a repurposed an abandoned petrol pump and pit stop. **[2]**

1

2

Experiences

DUBAI TURTLE REHABILITATION PROJECT, UAE

Run in partnership with Dubai's Wildlife Protection Office, this project has overseen the release of over 550 rescued sea turtles. Based at Burj Al Jumeirah, the centre nurtures rescued turtles back to health in a controlled environment. Satellites then track the progress of those released. It provides vital education for visitors and locals alike. *[1]*

SIRAJ CENTRE, PALESTINE

This award-winning not-for-profit creates links between Palestinian people and visitors via educational tourism, culture and homestays. Special events, like the Annual Olive Harvest and a Christmas Peace Pilgrimage, welcome people from all faiths and backgrounds. The centre also organizes a series of guided biking and hiking events.

ROYAL SOCIETY FOR THE CONSERVATION OF NATURE, JORDAN

Jordan is a pioneer in conservation-led tourism thanks to the work of the Royal Society for the Conservation of Nature and its Wild Jordan eco-tourism initiative. The RSCN has implemented 100 per cent local employment throughout its protected areas, meaning that eco-tourism supports over 160,000 families.

IRAQ AL-AMIR WOMEN'S COOPERATIVE, JORDAN

Managed and run by local women, but funded and set up by The Travel Corporation and Tourism Cares, the Iraq Al-Amir Women's Cooperative upskills unemployed women around remote Wadi Seer. Training includes how to make and sell handicrafts and pottery, and soft catering. Visitors can purchase goods at the onsite gift shop.

GILEBOOM AGRITOURISM, IRAN

A little ecolodge with a big heart, Gileboom Homestay has a whole host of experiences to get visitors closer to the surrounding countryside and villages. During Spring, when the hills come alive with wild flowers and the scent of orange blossom, guests can take part in day-long 'agritourism' experiences including tea processing, orange picking and rice-planting. *[2]*

3

4

Getting Around

EVRT DUBAI

Showcasing the slickest electric vehicles and tech, this five-day, 2,000-km, 100-per-cent electric adventure is billed as the 'world's most exciting Electric Vehicle Road Trip'. Seven car brands are taking part, and activities along the way will showcase the best of UAE. The aim of the trip is to accelerate the transition to 'smarter mobility'.

TRANSASIA EXPRESS

Carving between dynamic cities, colossal mountains and wooded river valleys between Istanbul and Tehran by train is one of the world's ultimate slow travel adventures. The 2,900-km route is broken down into several sections – a fast train, an overnight train, a ferry and another overnight train – so there's no chance of getting bored.

LEBANON MOUNTAIN TRAIL

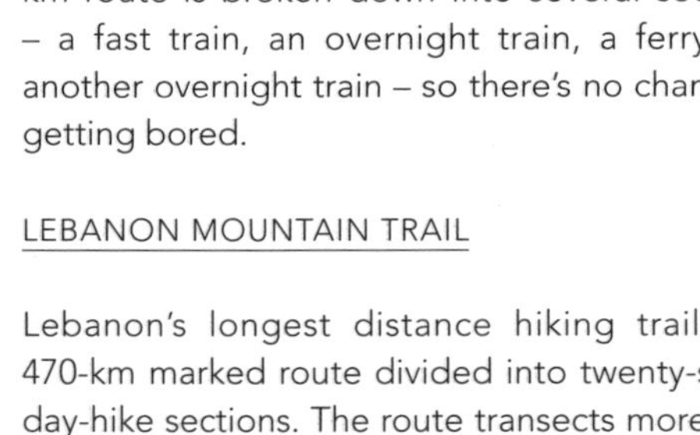

Lebanon's longest distance hiking trail is a 470-km marked route divided into twenty-seven day-hike sections. The route transects more than seventy-six settlements and from 570 m climbs to 2,073 m. The premise is to foster collaboration between rural communities, showcase Lebanon's unique heritage and landscapes, and provide economic opportunities.

FERRY FROM JORDAN TO EGYPT

Not one for those that like to avoid bureaucracy, this hour (or four hours on the slower car-ferry) ferry is none-the-less a far more exciting way to get across the Red Sea than jumping on a plane. ABMaritime runs boats between Aquaba and Nuweiba every day, chugging back and forth in front of Saudi Arabia's towering mountains. *[3]*

TOUR OF OMAN

Kicking off in Muscat, the Tour of Oman sees pro cyclists climbing mountains, skirting the coast, and weaving through cities. With smooth roads, a dry climate and varied landscape, many cyclists consider Oman to be a top destination. Whether you make like Chris Froome or take things more gently, Muscat has plenty of bike rentals to get you started. *[4]*

Asia

Whether independent backpackers or luxury retreat enthusiasts, travellers have flocked to Asia in pursuit of eastern philosophy, easy-going backwaters, frenetic cities and mountain kingdoms for decades. Beyond well-trodden trails through South East Asia, there's the lesser-explored highlands and heritage of Pakistan, Malaysia and Borneo, the Silk Road's landlocked Stans, Japan's unique temples and delicacies, and the world's allusive superpowers of Russia and China.

Asia is so vast and varied that it's a land of contradictions: home to the world's happiest country (Bhutan) and the world's most oppressed (North Korea); the birthplace of spiritual enlightenment (Buddhism) and of fast-consumerism (China). It's geographically diverse, too. Asia includes seven of the world's biodiversity hotspots, and the south-eastern region contains 20 per cent of the planet's species. Heading northwards, the Hengduan Mountain area of China is the richest temperate ecosystem in the world, and Siberia is one of the world's best-preserved wilderness areas.

With a burgeoning economy, expanding population, and conditions ripe for intensive palm oil production, habitat-loss is one of Asia's most pressing environmental concerns. The region has the highest rate of deforestation in the world, and consequently megafauna like elephants, tigers and orangutan are endangered. Global pollution is also felt acutely here; in piles of plastic and run-off from clothing factories. For anyone looking to travel sustainably in Asia, minimizing impact and supporting conservation efforts is vital.

Where to stay

TIGER MOUNTAIN LODGE, NEPAL

This pretty red-brick lodge, with unrivalled views across the Pokhara Valley, was built by hand involving over 300 local craftspeople. The property is virtually plastic-free, and everything is locally sourced. Its Community Support Partnership plays an active role in improving the surrounding area from funding forest rangers to primary school teachers. *[1]*

–

Over in Sri Lanka, Jetwing Hotels is another stalwart of sustainability; the family-run chain has been committed to waste and energy reduction for over forty years.

CARDAMON TENTED CAMP, CAMBODIA

Forming a unique partnership with Wildlife Alliance to protect Botum Sakor National Park, Cardamon's eight safari-style tents help to fund twelve forest rangers. The park has already seen a drop in poaching, hunting and illegal logging, which threatens the survival of species like pangolin. Guests can take part in conservation activities or sit back and soak up the jungle.

–

Not far away, on the raging Tmor Rung River, Shinti Mani Wild is pioneer designer and hotelier Bill Bensley's most sustainable property is equally at-one with the jungle.

AKAYRN, THAILAND

Following the opening of Asia's first single-use plastic free hotel, akra TAS Sukhumvit Bangkok, the Akayrn group have rolled the policy out across its entire portfolio. Impressively, this includes back of house; hotels work with suppliers to help them cut plastic from their supply chains. The hotel's Pure Blue Foundation educates locals and visitors, too.

–

Also in Bangkok, The Yard Hostel has waste top of mind; it's made from upcycled shipping containers and other reused materials. [2]

YASUESOU BIOHOTEL, JAPAN

Biohotel founder, Kazuhiko Nakaishi, took inspiration from the Austrian movement of the same name, and is on a mission to prove that a 100 per cent chemical-free, organic experience improves our well-being. The group's flagship property is Yasuesou in Nagano where bedding is organic cotton and scents are infused with locally grown camomile.

–

Fresh, organic ingredients, grown onsite or at nearby farms, is part of the winning formula at Nepal's top sustainable hospitality hero: Tiger Mountain Pokhara.

CEMPEDAK PRIVATE ISLAND, INDONESIA

Only a few hours overland and boat from Singapore, this newer, adult-only sister to Nikoi Island has the same commitment to sustainability based on The Long Run's 4Cs. Private villas made from local *alang alang* grass puncture the forest, blending seamlessly with the white sands as they tumble into an azure marine protected area. *[3]*

–

Equally idyllic and supportive of local communities is the laid-back resort of Wa Ale on an island in the heart of the Lampi Marine National Park in Myanmar.

MEI ZHANG, FOUNDER, WILDCHINA

WildChina shows travellers a different side of China, providing a sophisticated interpretation of local culture and nature through experiential, sustainable travel.

–

What's next for sustainable travel?

Making sure everyone knows it's accessible. Sustainable travel doesn't have to be more expensive (often it's cheaper) nor does it have to be centred around environmental conservation or ecolodges. Sustainable travel comes in many different shapes and colours; it's up to us to ensure people know what's available.

What gives you most hope for a more sustainable future?

The fact that we're talking about this. As more people discuss what sustainability means throughout different industries, we're moving ever closer to a time where sustainable practices are the norm, not an ideal to aspire to. Having this conversation (again and again) is the first step.

What advice can you offer those looking to travel more sustainably?

Get off the beaten path. Nothing says unsustainable more than the large-scale developments that shut out local communities and devastate the natural and traditional heritage of a place. Also, do your research to find out which hotels provide amenities that are free from single-use plastics or how to get a train to your next destination, rather than a plane. Finally, when you come across a community or organization working sustainably, invest. Stay at local hotels; visit wineries run by local families; buy gifts from women carrying on their cultural traditions.

1

2

Experiences

LEOPARD TRACKING, INDIA

From tigers to orangutans, pandas to snow leopards, Asia's wildlife encounters can be more allusive than those in Africa – there's less habituation and animals tend to roam inhospitable landscapes. One standout experience is the rewilding of 130 acres of leopard habitat at Sujan Jawai Leopard Camp among the craggy granite formations and sand river-beds of Rajasthan. *[1]*

HOMESTAY TREKS, BHUTAN, NEPAL AND INDIA

Some of the world's finest responsible travel initiatives exist high in The Himalyas where century-old paths between remote hillside villages offer the perfect combination of low-impact adventures and community-empowering homestays. Take a look at itineraries by Village Ways, Shakti Himalaya, Sasane Sisterhood, The Blue Yonder and Feheneh Travel.

TREE ALLIANCE, CAMBODIA

Sharing plates between temples in Siem Reap, local speciality crispy Mekong River weed in Luang Prabang, and family recipes in Phnom Penh; each one of these restaurants is run by TREE Alliance, a charity giving street kids a future. All profits from restaurants go straight back into training and social programmes for students in need. *[2]*

RIDING THE TRANSMONGOLIA TRAIL, MONGOLIA

There are endless horse-riding opportunities in Mongolia; choose small group adventures led by locals that stay in camps that leave no trace along the way. One of the most epic routes is the three-week TransMongolia Trail taking riders across The Steppe to the Khogn Tarna Sacred Mountain and volcanoes of Arkhangai.

HIKING NORTHWEST YUNNAN, CHINA

Home to Tiger Leaping Gorge and Shangri-La, this mountainous region hosts a dizzying array of legends, culture and nature of epic proportions. One of the most sustainable ways to experience it all is to hike the numerous mountain paths with locally owned and operated Lijang Xintuo Ecotourism Company, which employs Naxi and Yi ethnic minority guides.

Getting Around

WOMEN ON WHEELS, INDIA

Often surprising visitors and locals alike, this nimble social enterprise puts women from disadvantaged backgrounds behind the wheel in Dehli, Jaipur, Kolkata and Indore. The scheme ticks two boxes: it gives women a source of income and empowerment, and it gives female travellers greater peace of mind.

PILGRIMAGE ROUTES, JAPAN

With 1,000 years of Buddhist history unravelling among remote villages and pristine forests, the Kumano Kudo Pilgrimage is best taken slow. Along with the Camino de Santiago, it's the only other pilgrimage to receive UNESCO status thanks to the concentration of iconic sites throughout the beautiful Kii Peninsula. *[3]*

TRANS-MONGOLIAN, CHINA, MONGOLIA, RUSSIA

The Trans-Siberian is often cited as the East's most iconic train ride, but the Trans-Mongolian spans three cultures in half the time. The journey is sandwiched between equally mesmerizing but contrasting destinations: Beijing's sprawling temples, markets and high-rises, and vodka, omul and dachas on the edge of Lake Baikal.

THE REUNIFICATION EXPRESS, VIETNAM

The steel giants that ride along Vietnam's North–South Railway line are romanticized as the Reunification Express, because the tracks carve out a history of war, peace, and unity all at once. It can be an emotional ride. Trundling along the 1,726-km spine of Vietnam, through paddy fields, fishing villages and even city centres, it's the ultimate slow adventure.

SAILING PARAW EXPEDITION, THE PHILIPPINES

Jumping aboard The Batalik is an experience in itself; the traditional paraw is the largest of its kind and considered a relic. The boat was carefully resorted by the Paraw Expedition team, local artisans and historians. Beyond the boat, the itinerary itself is unique, taking guests into the remote clutches of forest-clad northern Palawan. *[4]*

Index

Acknowledgements

Thanks to Jeremy Smith, Megan Devenish, Richard Hammond, Graham Miller, Vicky Smith, Justin Francis, Fran Hughes, Delphine King, Anne-Kathrin Zschiegner, Joy Mbatau, Harriet Whiting, Francisca Kellett, Jane Anderson, Alice Callander, Alicia Brett, Naomi McKee, Anna Pollock, Susanne Becken, Julie Cheetham, the Long Runners, and many others for all the inspiration and support. Thanks also to so many wonderful hosts around the world; travel would be nothing without you.

Picture credits

2–3 Ljubaphoto/Getty Images; **6** Dylan Alcock/Shutterstock; **9** Cayuga Collection; **10** Rich Carey/Shutterstock; **13a** Innis McAllister/Alamy; **13b** Xuanhuongho/Shutterstock; **15** DimaBerlin/Shutterstock; **16l** Goodcat/Shutterstock; **16r** Olga Kashubin/Shutterstock; **17** United Nations Sustainable Development Goals (https://www.un.org/sustainabledevelopment); **18** FloridaStock/Shutterstock; **21** Anton_Ivanov/Shutterstock; **23** Kletr/Shutterstock; **24l** Karnizz/Shutterstock; **24r** Anna Krasnopeeva/Shutterstock; **25** Artem Fedorets/Shutterstock; **26** Everst/Shutterstock; **27** DimaBerlin/Shutterstock; **28a** Mumemories/Shutterstock; **28b** HildaWeges Photography/Shutterstock; **29l** Pumidol/Shutterstock; **29r** Magdanatka/Shutterstock; **30** Nazri Yaakub/Shutterstock; **32** Nguyen Quang Ngoc Tonkin/Shutterstock; **35** Julian Schlaen/Shutterstock; **36** STR/Stringer/Getty Images; **38** Fabian Plock/Shutterstock; **39** Sergey Uryadnikov/Shutterstock; **40l** Prosign/Shutterstock; **40r** Travelpix/Shutterstock; **43a** JoshoJosho/Shutterstock; **43b** Siderath/Shutterstock; **44** Chrisontour84/Shutterstock; **45l** Anansing/Shutterstock; **45r** Stanislava Karagyozova/Shutterstock; **46** CA Irene Lorenz/Shutterstock; **48** Mazur Travel/Shutterstock; **49** CharlesEvans/Shutterstock; **50** CloudVisual/Shutterstock; **52** Westend61 GmbH/Alamy; **55** Westend61 GmbH/Alamy; **56** Ymgerman/Shutterstock; **57** Alberto Loyo/Shutterstock; **59** Cayuga Collection; **60** Antony Souter/Alamy; **62** Good Hotel London; **63** Rafal Cichawa/Shutterstock; **64** R.A. Chalmers Photography/Alamy; **65** Oskar Hellebaut/Shutterstock; **67** Ira Shpiller/Shutterstock; **69** Eat Like A Local; **70** Image Professionals GmbH/Alamy; **73a** Global Himalayan Expedition; **73b** Samara Heisz/Alamy; **76** Fakrul Jamil/Shutterstock; **78** Bhanu Pratap Singh Rathore; **79** Kasia Nowak/Alamy; **80** Comuna do Ibitipoca; **81** Aardvark/Alamy; **83** Wolgang Kaehler/Alamy; **84** Lloyd Vas/Alamy; **85** Demamiel62/Shutterstock; **87** Jamie Lafferty/Guyana Tourism Authority; **89** Niebrugge Images/Alamy; **90** Kim David/Shutterstock; **92a** Colin Palmer Photography/Alamy; **92b** Moiz Husein/Alamy; **95** Namhwi Kim/Alamy; **96** Robert Harding/Alamy; **97** Alamy; **98** Juergen Ritterbach/Alamy; **99l** Minden Pictures/Alamy; **99r** Nature Picture Library/Alamy; **100** Cultura Creative RF/Alamy; **101a** Phil Dunne/Alamy; **101b** Nature Picture Library/Alamy; **103** Song_about_summer/Shutterstock; **105a** 3 Rivers Eco Lodge; **105b** Bozulek/Shutterstock; **106** Zbigniew Dziok/Alamy; **107** AgaDetka/Alamy; **108** Oliver Wintzen/Alamy; **110** Antony McAulay/Shutterstock; **111** TK Kurikawa/Shutterstock; **113** EQRoy/Shutterstock; **114** RafaGalvez41/Shutterstock; **116** Education Images/Getty Images; **119al** Hilke Maunder/Alamy; **119ar** Galumphing Galah/Shutterstock; **119bl** Karel Stipek/Shutterstock; **119br** AFP/Stringer/Getty Images; **120l** FiledImage/Shutterstock; **120r** ImageBroker/Alamy; **121l** Douglas Lander/Alamy; **121r** Bildagentur-online/Schickert/Alamy; **123** Yorkshireknight/Shutterstock; **124a** P. Spiro/Alamy; **124b** Hemis/Alamy; **126l** Jit Lim/Alamy; **126r** Robert Mutch/Shutterstock; **127l** David Hunter/Alamy; **127r** Wolfgang Kaehler/Getty Images; **128** Guaxinim/Shutterstock; **131al** Kelly Venorim/Shutterstock; **131ar** K.D. Leperi/Alamy; **131bl** Bettina Strenske/Alamy; **131br** Alison Wright/Alamy; **132l** Alan Falcony/Shutterstock; **132r** Wolfgang Kaehler/Getty Images; **135** Supergenijalac/Shutterstock; **136a** Kolarbyn Ecolodge; **136b** Building Image/Alamy; **138l** Mark Baynes/Alamy; **138r** Nature Picture Library/Alamy; **139l** Varuna/Shutterstock; **139r** Phaustov/Shutterstock; **140** Robert Harding/Alamy; **143a** Ariadne Van Zandbergen/Alamy; **143b** Martin Harvey/Alamy; **144l** Education Images/Getty Images; **144r** Gallo Images/Alamy; **145l** Arterra/Getty Images; **145r** Mauritius Images GmbH/Alamy; **147** GTW/Shutterstock; **148a** Juergen Hasenkopf/Alamy; **148b** Eric Lafforgue/Alamy; **150l** National Geographic Image Collection/Alamy; **150r** Ian Bottle/Alamy; **151l** Art Directors & TRIP/Alamy; **151r** Justin Setterfield/Getty Images; **152** Wattanal/Shutterstock; **155a** Alamy; **155m** The Yard Bangkok; **155b** Moonie's World Photography/Alamy; **156l** Bhasmang Mehta/Shutterstock; **156r** Hemis/Alamy; **157l** Chrytite RF/Alamy; **157r** Michael Wels/Alamy

First published in 2021 by White Lion Publishing,
an imprint of The Quarto Group.
The Old Brewery, 6 Blundell Street,
London, N7 9BH,
United Kingdom
T (0)20 7700 6700
www.QuartoKnows.com

A catalogue record for this book is available from the British Library.

ISBN 978-0-7112-5601-9
eISBN 978-0-7112-5602-6

10 9 8 7 6 5 4 3 2 1

Designed by Sarah Pyke
Printed in China